CW00765467

OUT OF THIS WORLD
AND INTO THE NEXT

OUT OF THIS WORLD AND INTO THE NEXT

Notes from a Physicist on Space Exploration

ADRIANA MARAIS

Profile Books

First published in Great Britain in 2025 by
Profile Books Ltd
29 Cloth Fair
London
EC1A 7JQ
www.profilebooks.com

1 3 5 7 9 10 8 6 4 2

Typeset in Sabon by MacGuru Ltd
Printed and bound in Great Britain by
CPI Group (UK) Ltd, Croydon CR0 4YY

A CIP catalogue record for this book is available from the British Library.

We make every effort to make sure our products are safe for the purpose
for which they are intended. For more information check our website
or contact Authorised Rep Compliance Ltd., Ground Floor, 71 Lower
Baggot Street, Dublin, D02 P593, Ireland, www.arccompliance.com

ISBN 978 1 80081 980 1
eISBN 978 1 80081 983 2

ACKNOWLEDGEMENTS

I feel like this book has written itself, so my heartfelt thanks and appreciation go out to all people who have dedicated their lives to contributing to human knowledge, and to everyone I've known for our cherished interactions that have shaped the thoughts compiled here. Especially my parents, siblings, friends, partners and mentors, who have deeply influenced the way I see the world. I am grateful for having had the opportunity to share my ideas with hundreds of audiences on all continents of our planet and I thank those of you who have challenged and broadened my thinking. We can live in harmony with each other and the environment, wherever in the Universe we may be. Thank you to our Earth for having put up with us for so long while we learn how.

CONTENTS

Part II: Who are we?

Part III: Where are we going?

PREFACE

*'A species far-wandering and equipped for distant
migrations, through inborn wanderlust or otherwise,
would always have a better chance than one confined.'*
Eugène Marais (1871–1936)

When I first read about the call for volunteers for a one-
way mission to Mars in 2012, I felt suddenly nauseous,
struck by a childhood memory that rushed back to me
with perfect clarity: as my black plastic scooter rattled
down the brick driveway at playgroup one day, out of
the blue I had a series of imaginings. There was a global
radio broadcast (this was the pre-Internet 1980s): a call
for a volunteer to go on an urgent journey to find a new
home for humanity. The volunteer would travel through
space far away from Earth, and send back a message if a
suitable planet was discovered. I vividly pictured myself
looking out of a small round window with a notebook
in hand. The distant stars were stationary although I
was travelling fast. It was dark and I was alone. I hoped
for the sight of land, the welcome arc of a planetary

horizon, but knew that I may not live long enough to get there. While this time I turned round to kick back up the hill on my scooter, I silently vowed that I would volunteer for such a critical mission for humanity, even if I may not come back. I must have been four or five. It would be more than twenty years until I and people all around the planet would be confronted by this same scenario: would you volunteer to leave Earth to establish the first off-world society?

When most people think about 'home', an area much smaller than the surface of Earth comes to mind. For many, Earth is their favourite planet. But for those who feel a curiosity, an affinity and indeed a sense of belonging with that overwhelming majority of what is beyond, Earth is but a pale blue dot in a Universe of star stuff waiting to be known. Here, I'd like to share my wonder at being alive at this extraordinary time on Earth. Four billion years of evolution on this planet have brought us to the brink of a new era: just decades since we first went to space, it won't be much longer before we're building new worlds beyond home.

PART I

WHERE DO WE COME FROM?

One night, at a big space conference, a friend and I decided to sneak into a prestigious invite-only dinner (to which we were not invited). It was 2016 and we knew that the founder of SpaceX, Elon Musk, would be presenting his plans to make humanity multiplanetary on the main stage the following day. Who knew whom we might meet there. From the door we identified two empty seats, strode briskly past security and sat down as casually as possible. I nonchalantly took a large gulp of what I thought might be juice in front of me. It turned out to be tequila – we were after all in Mexico. The man next to me looked impressed that I had swallowed the whole lot, and, skipping the small talk, asked what I felt when I look out into the stars.

'A sense of belonging,' I replied without hesitation. The man was a cosmonaut, and a discussion about our place in the Universe ensued which has remained deeply etched into my memory to this day.

As children, our homes are the houses we grow up in, surrounded by the people and places that become familiar to us. As we grow up, our perception of home grows with us, and as we learn about space and time and causality – we may not frame it in those words – a question every child asks is: where do we come from? While we have all wondered about this at some point, a few of us spend lifetimes searching for answers. The quest to understand where we come from, to understand our place, our home, in the Universe is driven by an innate curiosity about the reality beyond our world, by our imagination of what lies beyond our current experience.

Did you know that 10 percent of our bodies is as old as the Universe itself? According to our best theory of how it started, anyway – the Big Bang Theory – and the rest of us is made up of the dust of exploded stars. To understand our origins, our place in the Cosmos in which we find ourselves, we must follow the allure of the unknown beyond Earth, back in time to where it all, and we all, began, on a timeline beginning billions of years ago.

1

THE UNIVERSE

In the beginning: the Big Bang

In the broadest sense, the place that we are from is the Universe. And, according to current scientific theory anyway, that is where we will remain, since while space is vast from our human perspective, there is in fact a boundary to reality: the edge of the Universe, beyond which no travel, communication or any kind of interaction is possible.

As the story goes, in the beginning, around 14 billion years ago, the entire Universe burst out from an infinitesimally small point, in an event first jokingly referred to as the 'Big Bang'; the name, however, has stuck. Over the next few millionths of a second, the Universe rapidly expanded and was filled with an incredibly hot and dense plasma – a fourth state of matter not solid, liquid or gas – made up of the fundamental building blocks of matter called quarks as well as negatively charged particles called electrons. Within the first second, the quark soup cooled enough to form matter as we know it, including protons and neutrons – the particles that together make up the nuclei of atoms.

A full-scale simulation to test our theory of the Big Bang requires more energy than we can conceive of controlling. However, the world's largest particle accelerator, the Large Hadron Collider at CERN, smashes things like the nuclei of atoms together at extremely high energies, forming something akin to the quark soup thought to have existed shortly after the Big Bang. Observing these mini big bangs gives us some clues about what was happening in the early Universe.

We've learned that as the Universe continued to expand and the temperature and the density decreased, some 380,000 years after the Big Bang, protons and neutrons combined into positively charged nuclei and attracted the free negatively charged electrons roaming around to form the simplest atoms, hydrogen and helium. As these atoms formed, excess energy was released in the form of photons – the particles that make up light – and with the free electrons out of the way, these photons of light could travel uninterrupted. Before this, the Universe was opaque. And then it became transparent as the oldest light in the Universe streamed out. Awe-inspiringly, the now-cooled remnants of this light are still observable today in the form of the Cosmic Microwave Background.

This is as far back as we can see, as these primordial photons arrive at our planet after travelling through the expanding Universe for billions of years. We only noticed this backdrop to our Universe around fifty years ago. At first mistaken for interference from pigeon droppings on the equipment, the Cosmic Microwave Background was accidently detected by astronomers Robert Wilson and Arno Penzias. After the pigeons were removed and the new supersensitive antenna cleaned, they realised that the noise they were picking up was actually a signal, and that the space in between stars and galaxies was not completely dark, as was previously believed. In the background, coming from all directions at all times, is a faint signal: an ancient bath of light cooled to just slightly above absolute zero, left over from the formation of matter in the Universe billions of years ago.

Coupled with our observations that everything is moving away from us and that most galaxies are moving away from each other, faster and faster, implying that the Universe is expanding at an accelerated rate, analysis of the Cosmic Microwave Background lets us wind back the clock. By doing this, we can estimate the age of the Universe: somewhere in the region of 13.8 billion years.

Looking at the oldest light tells us not just about when the Universe came into existence, but also about the earliest matter: after 380,000 years the Universe consisted of mostly hydrogen and some helium atoms. In fact, the Big Bang is the only process we know of that produces hydrogen in significant amounts in space. Therefore, the water molecules making up more than half of your body contain hydrogen atoms (constituting around a tenth of your mass) that are almost 14 billion years old!

Though the beginning of the Universe may feel so very far away, there is something magical in the knowledge that we are all bathed in faint, ancient light from the beginning of time. That within each of us lives the entire history of our Universe.

After the formation of the first hydrogen and helium atoms, clumps of this early matter began to gather under the force of gravity into hot, bright, dense objects: the earliest stars were born. Within the nuclei of the atoms that make up matter, vast amounts of energy are stored; under the extreme temperatures and pressures in their cores caused by the immense gravity there, stars release this nuclear energy in a process called fusion, which results in hydrogen nuclei being fused into helium nuclei. The mass of the helium produced is less than the two input hydrogens, and according to the revelation that energy and matter are interchangeable, as first described by physicist Albert Einstein's famous equation more than a century ago, the difference comes streaming out in the form of light as well as particles.

Gradually, structures of matter of increasing size began to emerge, and after just a few hundred million years collections of early stars formed galaxies. Once a star has burned all of its hydrogen it begins to die. While most stars eventually fade away, some meet their ends in spectacular explosions called supernovae. The energy released during the lifetime of stars and during such supernovae events results in the formation of the rest of the elements comprising the periodic table: the oxygen, carbon, nitrogen and so on, which make up all matter, including our bodies. As astronomer Carl Sagan said, we are made of star stuff.

The Big Bang is our best-supported theory of how the Universe began: with an ancient explosive force forming a reality expanding at increasing speeds in all directions. Tempering any hubris, however, I think we can agree that some significant unresolved issues in our understanding of the Cosmos remain.

Looking back at the history of the Universe in this way has led us to the somewhat awkward conclusion that the majority of what is out there – a whopping 95 percent – remains hidden from sight. If our current theories are accurate and the rate of expansion of our Universe is continuing to accelerate over time, then there must be a mysterious, so far unseen force causing this acceleration. We call this dark energy. Furthermore, observations of objects in the Universe, for example galaxies rotating at such high speeds that their gravity shouldn't be able to hold them together, don't make sense unless a large amount of unseen stuff – dark matter – is also playing a role. We have not yet observed dark matter directly, precisely because it is 'dark'; it doesn't interact with light or matter as we know it, making it impossible to detect with existing instruments. Its existence is therefore inferred from such observed anomalies. More recent results may indicate that dark energy itself is evolving with time, once again flummoxing our understanding of cosmology.

Another challenge in understanding the Universe is that we have not yet managed to unify our theory of reality on the very small scales where particles like quarks, electrons and photons exist – quantum mechanics – with our theory of space and time in a Universe containing large quantities of the very small – general relativity.

Quantum theory tells us some strange things, some of which we'll get to later, but basically that what we observe in the Universe consists of quanta, or units that can't be broken up further: light is made up of indivisible particles called photons; electric current, of electrons; and protons and neutrons are made up of quarks (all of which do indeed so far seem to be indivisible). And these particles don't behave like things we are used to: they can manifest as particles or as waves with frequencies and wavelengths. Furthermore, fundamental particles with a shared history – like the Big Bang – can retain special quantum correlations appearing to transcend our understanding of space and time: in what Einstein called 'spooky action at a distance', also termed entanglement, what happens to one impacts the other seemingly instantly no matter how far apart they are. While we don't yet know if the space and time in which these quantum objects exist are also quantised, these quanta certainly don't age like we do, moving equally forwards or backwards in time according to the quantum mechanical equations that describe them.

In the theory of general relativity, space and time are continuous, combined into a single concept called spacetime; and things with mass bend spacetime. In a memorable physics demonstration, Tony Fairall, one of my astronomy lecturers, wrapped a wooden frame in cling film, this surface being a two-dimensional analogy to four-dimensional spacetime. He placed a large marble in the centre of the surface, which caused the cling film to sag in the middle. He then placed a smaller marble at the edge of the surface, gave it a push, and we watched as it circled inwards towards the big marble. Similarly, a star (the big marble) curves the spacetime round it, such that a planet (the small marble) in the vicinity, given some initial velocity, begins to orbit the star; without the friction of the cling film, this orbiting continues for a long time.

The idea that mass bends spacetime to produce gravity, with the result that things in the neighbourhood travel in curves, was rather spectacularly demonstrated just a few years after Einstein proposed the theory in 1915. If Einstein's theory were accurate, the light from stars positioned behind the Sun as seen from Earth should be bent by its gravity and they should therefore appear to be in different positions from where they actually are. An eclipse, where the Moon blocks enough sunlight for stars near the Sun to be seen from the surface of Earth, was an ideal opportunity to test this. And indeed, during the total solar eclipse of 1919, measurements showed that Einstein's predictions were correct: the locations of the stars now visible appeared displaced compared with their actual positions because the Sun was in between and bending the spacetime, and thus the light travelling between the stars and Earth.

Without going into too much of the detail developed by thousands of physicists over the past few hundred years, as a result of the inconsistencies between these theories, in particular in our understanding of time, some conundrums remain. For example, it's difficult to say whether there are in fact 10 to the power of 80 electrons (1 followed by 80 zeros) or just one electron in the Universe. Physicist Richard Feynman said he received a telephone call one day from fellow physicist John Wheeler:

'Feynman, I know why all electrons have the same charge and the same mass.'

'Why?' asked Feynman.

'Because, they are all the same electron moving backwards and forwards in time!'

It is worth mentioning at this point that while we have casually given times in units like years here, our understanding of time is limited by our inability to reconcile quantum theory with general relativity: time in quantum theory is absolute, steadily ticking away at all places in the Universe, whereas in general relativity it is dynamical, since gravity can bend spacetime and therefore time itself. We can, however, define cosmic time as a time that all observers co-moving with the expansion of the Universe agree upon. Then we can say that the Universe is 13.8 billion years old. Neither theory, unfortunately, has much to say about what happened before the Universe burst out, as spacetime, and therefore the physics to describe it, did not yet exist. As physicist Stephen Hawking put it: 'Asking what came before the Big Bang is meaningless, because there is no notion of time available to refer to. It would be like asking what lies south of the South Pole.'

While we are yet to understand why the Big Bang happened, how space and time emerged from this event, and why, of all the physical possibilities, our Universe is one in which stable forms of matter and therefore life can evolve, the stage that we call reality was set. Leaving these mysteries aside for now, let's return to the things we can see: the estimated 5 percent of the Universe that is made of light and matter as we know it, and the spectacular structures that have emerged in the past 14 billion years – the things that have made life on Earth possible.

What we can see: galaxies, stars and exoplanets

So, what has happened in the 14 billion years since the Universe was born? What does it look like now? And where do we fit into all of this?

A galaxy is an immense collection of gas, dust and billions of stars as well as their planetary systems. Galaxies contrast massively with intergalactic space, which is comparatively empty. The stars that make up a galaxy evolve over time, and when they eventually die black holes sometimes form. The explosive supernova that occurs when a large star collapses under the force of its own gravity – typically releasing in just a few seconds as much energy as our Sun does over its entire lifetime – can leave behind a black hole with a mass that can range from a few to hundreds of times the Sun's mass. A black hole is a region where gravity is so strong that nothing, including light, has enough energy to escape. Large galaxies tend to have supermassive black holes at their centres, which can range from 100,000 to even billions of Solar masses. While there's a lot we don't know about supermassive black holes, images of our galactic centre and the supermassive black hole there produced by the Event Horizon telescope and the MeerKAT precursor to the Square Kilometre Array radio telescope continue to shed light on the issue.

Of the trillions of galaxies that have formed in the Universe, all the stars that we can see with our naked eyes in the night sky are in fact in our Galaxy, the Milky Way. The Milky Way is a rather old neighbourhood, relatively speaking, containing at least 100 billion stars, some of which are also the oldest known stars in the Universe. Astronomers determine the age of stars by examining groups assumed to have been formed at similar times. By observing their brightness and temperature and comparing this to models of what stars look like at various points in their evolution, we can estimate the ages of stars. The so-called 'Methuselah Star', named after the longest-lived character in the Bible, is located around 200 light years away from Earth in the constellation of Libra. Methuselah is estimated to have formed soon after the

Big Bang and is one of the oldest stars we know of. Our star, the Sun, is a newcomer by comparison at around 4.6 billion years old.

From our vantage point near the middle of the Galaxy, some 25,000 light years away from both the edge and from the supermassive black hole at the centre, we have very little direct experience of what is out there. Gazing up at the starry night sky, more recently with increasingly sophisticated telescopes to analyse the incoming light signals, we get a sense of infinite possibilities. Are there stars like our Sun out there? Do they also have planets? Could life have emerged there?

Stars shine, making the first question easier to answer: around 20 percent of stars in the Milky Way are Sun-like, meaning that they have near-solar surface temperature, size, chemical composition and age. Planets are much more difficult to detect; and only recently (in 1995) was the existence confirmed of the first planet around a solar-type star beyond our Sun, a so-called exoplanet – by physicist Didier Queloz, a PhD student at the time, and his supervisor Michel Mayor, both of whom won the Nobel Prize in Physics in 2019 for the discovery. Exoplanets don't reflect enough light to be seen directly from Earth. They are instead detected indirectly, for example by observing the wobbling of the star's centre of mass due to a planet's presence, observable in the form of minute changes in the frequency of the star's light. Since that first discovery, thousands of exoplanets have been identified. The oldest known exoplanet, named the Genesis

Planet, is located around 12,400 light years from Earth in the constellation of Scorpius. Genesis is believed to be about 12.7 billion years old; almost three times as old as Earth. Even our nearest exoplanets orbit a star more than four light years away. In sum, our understanding of the ground conditions on planets beyond our Solar System is indeed limited.

While the nearest exoplanets remain beyond our reach, we have sent technology out of the Solar System: in the 1970s the Voyager missions were launched, carrying gold phonograph records engraved with messages from 'a small, distant world, a token of our sounds, our science, our images, our music, our thoughts and our feelings', as described by US president Jimmy Carter at the time. After visiting multiple planets in our own Solar System, these craft became our first technology to venture out of the Solar System into the vast darkness of interstellar space, the space between star systems within a galaxy. Of the three stars in the nearest system called Proxima, Proxima Centauri – first observed from Johannesburg – is the closest. However, at Voyager 1's current speed of over 60,000 kilometres per hour, it is more than 70,000 years away. Even at 300,000 kilometres per second, light takes more than four years to travel there.

This has not stopped big ideas of getting lightweight, fast-moving technology over there in our lifetime to have a look. The Breakthrough Starshot project aims to use the most powerful laser array ever built on a light sail

that would accelerate a tiny craft with a mass of around
a gram to 20 percent the speed of light. A journey to
Proxima Centauri at that speed would take about twenty
years. Add another four years for a light signal to return
to Earth, and that means we could still be alive to receive
the first data from another star system. What we will
find there is anyone's guess.

Proxima Centauri b is the closest exoplanet, discovered in 2016 orbiting in the habitable zone of Proxima
Centauri. Since all known living organisms require water,
for a planet to be in the habitable zone means it orbits
a star at a distance where liquid water could exist on
its surface, given a dense enough atmosphere. Proxima
Centauri b has a mass slightly greater than Earth's, and
orbits its star at a distance around twenty times closer to
the Sun than Earth, with a year of approximately eleven
Earth days. So close to Proxima Centauri, fast-moving
flows of charged particles ejected from the star, called
solar winds, could be thousands of times more intense
than on the surface of Earth; and although the planet
may host liquid water, under these conditions its ability
to support life is unknown, and will likely remain so
until a mission like Starshot can provide us with new
information.

Can we take Starshot thinking even further, though?
If we can envisage sending our technology to a planet in
the habitable zone, what else could we also send there? At
a space congress I attended, one memorable speaker proposed that a tiny cargo be added to a Starshot-like craft.

The cargo, terrestrial bacteria, could then be delivered to a suitable planet in the Proxima System to seed life there for study. At the time, from the back of the small lecture room, I found the thought of this hilarious. If we were being featured in an intergalactic reality entertainment channel, the commentary might have gone along these lines: 'In this, the 4 billionth year of the Milky Way's *How to Grow Intelligence from Scratch* saga, the subjects of the experiment on planet Earth propose to run a similar experiment of their own. A pivotal moment for the beings hailing from the ancient Methuselah System in the Libra Constellation who have been running such experiments, so far without success, for most of the age of the Universe. For the first time, the life they seeded on Earth is planning interstellar travel. But will the human society prevail for long enough to get there before the statistically likely self-annihilation we have seen in countless previous episodes? Stay tuned for the season of the Eon!'

But seriously, we estimate that there are more than 40 billion Sun-like stars in our Galaxy, each with, on average, at least one planet in orbit. If we try to imagine what terrestrial life would be like after 10 billion more years of evolution, we can start to understand that if life emerged in the early Universe, for example around the Methuselah Star or on the Genesis Planet, it would likely be completely unrecognisable to us. Also, we might realise that intelligent beings out there may have had this same idea some time ago – that it would be interesting to introduce life to nearby planets, to see what happens. It is not impossible that what I had just witnessed was really the subject of a much older life-seeding experiment here on Earth unwittingly presenting a proposal to run his own similar experiment elsewhere. There is a theory of the origins of life called panspermia, proposing that primordial life was delivered to Earth from space billions of years ago. But we're getting ahead of ourselves.

To explore beyond our Solar System, even remotely, will require new propulsion technology. At current maximum rocket speeds, it would take tens of thousands of years to get to the nearest exoplanet, Proxima Centauri. A revolution in physics and our understanding of spacetime will be required to realise interstellar travel, never mind visiting other galaxies, which are millions and millions of times further away. While the Universe is vast, luckily for us there are some rather interesting places that may shed light on our quest to understand our origins which are not so far from home.

Places we can observe: our Solar System

Our Sun is at the heart of things here in our Solar System. In the powerful gravity of our newly formed star, swirling masses of debris gathered gradually and often violently into the orbital bodies making up our System. In addition to its gravitational influence due to its mass – which makes up an impressive 99.8 percent of the total mass in our Solar System – the Sun's heat and light have enabled a fascinating range of events. Including, on at least one orbital body, the emergence of life.

Our Solar System is an island of activity in the vastness of interstellar space. Although it is our locale, so to speak, and relatively close – so much so that parts of it are visible to the naked eye – our knowledge of it is mostly inferred. For the vast majority of our time on Earth, we have been limited to observing the night sky from our vantage point on the surface of our planet, as well as picking up rocks that have fallen from space. Our first direct knowledge of any place other than Earth was through crewed visits to our Moon half a century ago. More recently, our progress in robotics and automation has opened up exciting avenues for learning about the Solar System via remote investigation, and we have brought back samples from the solar wind, both sides of the Moon, multiple asteroids and a comet for further analysis. While things get more extreme the further out we go – from our Earthly perspective at any rate – some worlds may yet hold surprises for us in our quest to understand the origins of life back home.

Earth is the only planet we know of to host liquid water on its surface; the oceans that cover most of it are one of its most striking features. Our oceans are abundant with life: just a drop of seawater can contain over a million different species of microorganisms. The search for water has driven much exploration in our System; both in terms of the possibility of finding life or evidence of life and the potential for establishing life beyond

Earth. In fact, this molecule so vital to terrestrial life has been detected in a range of other places in the Solar System, perhaps unsurprisingly, as hydrogen and oxygen (along with helium) are in the top three most abundant elements in the Universe. The abundance of the H_2O molecule in off-world locations in our Solar System is exciting in terms of understanding the history and future of life in our System, leading naturally to questions like: are these strange worlds places we could visit? Might we find clues to our origins, or even life there?

The Solar System is typically divided into two regions: the Outer System and the Inner System. Our direct knowledge of the Outer System is as limited as the faint sunlight there – only about one-thousandth of the light received by our planet reaches the furthest planet, Neptune. The icy outer reaches beyond Neptune are the last frontier of our System, potentially holding pristine clues to the origins of the planets and perhaps even ourselves (out here ice could be frozen water, methane, ammonia or other compounds). Though part of our System, it's easy to forget how little we know – still – of this furthest realm.

Neptune's existence was predicted in the 1840s by mathematician Urbain Le Verrier based on observations of the movements of the known planet Uranus. By analysing small perturbations in its orbit, Le Verrier concluded that an as yet undetected planet of a similar size to Uranus was orbiting just beyond it. On receiving a letter from Le Verrier, astronomer Johann Galle discovered Neptune that same night, at a location just a degree off from Le Verrier's prediction. Subsequent to this spectacular demonstration of the predictive power of theoretical physics, what lies beyond Neptune in the cold, dark expanse of the Outer System remained a mystery for almost a century.

Then, in 1930, further anomalies in the orbit of Uranus attracted the attention of astronomer Clyde Tombaugh. While Pluto may have been downgraded from planet status in 2006 by a committee reasoning that it had not yet cleared enough neighbouring objects to qualify, the story of Tombaugh's discovery of Pluto did end rather fantastically with him becoming the first person to visit it (posthumously, that is). Besides carrying Tombaugh's ashes to Pluto, the New Horizons mission to the Outer System in 2015 revealed that the dwarf planet has five moons, the largest being about half its size, a thin atmosphere of mostly nitrogen, methane and carbon monoxide. While average surface temperatures are below negative 230 degrees Celsius, surprisingly, Pluto's surface features indicate geological activity, which, coupled with potential internal radioactivity, may be sufficient to support a subsurface ocean. If sustainable so far from the Sun, what may lurk within these waters remains a compelling mystery.

Beyond Pluto, the existence of a belt of objects in the far reaches of the Outer System was predicted in the 1950s by astronomer Gerard Kuiper, and was finally detected in 1992 by astronomers Dave Jewitt and Jane Luu, who had been scanning the skies in search of dim objects beyond Neptune for years. The Kuiper Belt comprises a vast ring of millions of icy bodies glinting dimly in the light of the distant Sun. Far enough away to have avoided the gravitational pull towards the larger planets, these Kuiper Belt bodies are remnants of the early Solar System and can provide valuable insights into its origins. Some of them are also behaving strangely.

In spite of the discovery of a range of icy worlds at the edge of the Outer System, there remains an elephant in the room: the irregular motion of a group of Kuiper Belt objects could be indicative of a planet with a mass of as much as ten times that of the Earth, and an elongated orbit hundreds of times as far from the Sun as the Earth. And we've seen how such orbital anomalies have given rise to the prediction and detection of large celestial bodies in the past. Could it be that there is a ninth planet in the Solar System (sorry Pluto), bigger than Earth, that we've not yet noticed?

Author Zecharia Sitchin claims that Sumerian mythology refers to a planet called Nibiru going round our Sun with a highly elliptic orbit of 3,600 years extending from the Asteroid Belt to over ten times further from the Sun than the dwarf planet Pluto. Sitchin's planet is typically dismissed by arguments of orbital stability – such close encounters with the planets in the Inner System would result in dramatic changes to the proposed orbit within just a few revolutions. However, the fact that we have not been able to disprove the existence of an as yet unde-tected planet larger than our own orbiting our Sun leaves us to conclude that our knowledge of even our own Solar System is decidedly limited.

The distant body Farfarout (yes, discovered beyond the previously furthest known body Farout) remains, for now, the most distant catalogued Kuiper Belt object in the Outer System. Farfarout is a dwarf planet about 400 kilometres across and around 20 billion kilometres from the Sun; even light takes nearly twenty hours to reach it. These distant frozen worlds, largely untouched by the turbulence out of which the bodies closer to the Sun formed, contain the building materials for our System in their primary form. How this icy realm may yet contrib-ute to our understanding of our System's origins, and perhaps our own, remains to be seen; the New Horizons mission currently in the Kuiper Belt is planned to be operational until the end of decade.

Moving a bit closer to home, some of the largest celestial bodies in our System are big and bright enough to be familiar features of our night sky. The nearer regions of the Outer Solar System are home to four giant planets that make up 99 percent of the mass of known objects orbiting the Sun. They are referred to as Jovian planets, after the largest of the four: Jupiter. More than ten times bigger than Earth and with gravity 2.5 times stronger, most of Jupiter's mass consists of hydrogen and helium. This means you would weigh more than double there than what you do on Earth, if there were any solid ground to stand on. Jupiter's gaseous surface surrounds an outer core likely of liquid metallic hydrogen producing its powerful magnetic field, and within that a dense inner core potentially of solid rock, metal and ice. Saturn is the second-largest planet in the Solar System, with a range of characteristics in common with Jupiter. Saturn's most famous feature is its ring system, its startling beauty having been captured in high resolution by the Cassini mission in 2017, just before the spacecraft's dramatic plunge into Saturn's atmosphere. The rings are mostly made up of ice particles, and also debris and dust, thought to be pieces of space rocks or even moons shattered by Saturn's powerful gravity before reaching the planet.

Slightly smaller than these two gas giants, the ice giants Uranus and Neptune are each around four times bigger than the Earth. Uranus' name refers to the Greek god of the sky, who according to mythology was the great-grandfather of Mars, the grandfather of Jupiter and father of Saturn. Uranus and Neptune are similar in composition to their larger siblings, with hydrogen-rich atmospheres below which is predominantly ice, thought to be made up of water, methane and ammonia, as well as rock. Uranus has the coldest atmosphere of all the planets in the Solar System, with a minimum temperature of negative 224 degrees Celsius. While the almost right-angled tilt of its equator to its orbital plane gives Uranus the most extreme seasons in the System, Neptune is home to some of the strongest winds, which reach speeds of over 2,000 kilometres per hour.

Jupiter, Saturn, Neptune and Uranus have surfaces of swirling gases; attempting to land on any of these gaseous and icy giants would result in being sucked down by a powerful gravitational field and then crushed by rapidly increasing pressure and density closer to their solid central cores. It's difficult to imagine what kind of life, if any, might originate in such foreign environments compared with what we are used to closer to the Sun. However, there are some worlds with surfaces in the Outer System that we might find more familiar: moons, hundreds of them, detected in orbit around the outer planets.

To support liquids on the surface of a celestial body, some kind of atmosphere is required. Titan, Saturn's largest moon and the second largest in the Solar System, is the only moon known to have a dense atmosphere, which, like Earth's, is made up of mostly nitrogen, also with small amounts of methane (though at a 50 percent higher pressure than ours). The opaque atmosphere prevented us from knowing much about the surface for a long time, until a lander was deployed from the Cassini mission to touch down there in 2005, revealing liquid lakes in the polar regions. Titan is the only known body besides Earth to have liquids on, as opposed to beneath, its surface. These liquids, however, are not water but hydrocarbons like methane and ethane, raining down from clouds and evaporating back up into the atmosphere at maximum temperatures of around negative 180 degrees Celsius. Beneath all this, Titan is also thought to have a subsurface ocean of liquid water and ammonia. Perhaps the Dragonfly mission, scheduled to depart from Earth later this decade and which will take seven years to reach Titan, will provide some answers.

The giant planets' huge masses, coupled with the highly elliptic orbits of some of their moons – meaning that these moons alternatively pass close to and also travel far from their planets during each orbit – result in powerful tidal forces that through friction may create enough heat to maintain liquid water beneath some moons' icy exteriors, even in the absence of a significant atmosphere. Some of these moons are also geologically active, and may contain radioactive material producing heat. From Pluto to the moons of the gas giants, there are in fact a range of subsurface environments in the Outer System where, even so far from the Sun, we believe there may be enough heat to support the presence of liquid water.

And in recent years, what appear to be plumes of water vapour have indeed been observed venting into space from the fissured crusts of both Jupiter's moon Europa and Saturn's moon Enceladus, revealing geological activity and potentially the presence of salty oceans below their frozen surfaces. Moreover, the precursors of amino acids, as well as phosphorous, necessary for life as we know it, were detected by Cassini in a vapour plume emitted by Enceladus in 2015. Ganymede, also a satellite of Jupiter, is the largest and most massive of all the Solar System's moons, and is the only moon known to generate a substantial magnetic field of its own. Recent detection of water vapour in its thin atmosphere indicates that it may contain a lot of water beneath its icy crust; possibly more than all Earth's oceans combined.

While these Outer System worlds remain years beyond Earth with current propulsion technology, the fact that we have strong indications of the presence of subsurface water has important implications for understanding our origins. As we await upcoming missions scheduled to collect further samples from these moons and their eruptions, there's little more we can do than imagine what, or who, may be dwelling beneath the ice … distant cousins of terrestrial life? Or, even more intriguingly, complete strangers?

Closer to home, in the region between the orbits of Mars and Jupiter separating the Inner and the Outer Solar System, we arrive at the Asteroid Belt; home to many smaller bodies orbiting our Sun. While water in solid form, ice, is a common component of asteroids, the existence of liquid water, which requires heat, was thought to be limited to larger bodies like moons that are geologically active and have a highly elliptic orbit around a massive planet. That is, until the Dawn mission of 2015 detected something interesting. Ceres is a rather large rock found in the Asteroid Belt; with a diameter of 1,000 kilometres, it is classified as a dwarf planet rather than an asteroid. On its surface we see evidence of cryovolcanic activity. As the name might suggest, a cryovolcano is a type of volcano that emits volatiles (substances that evaporate easily) like water or methane, which freeze on ejection into an extremely cold environment. Ceres is the closest known cryovolcanic body to the Sun, erupting roughly every 50 million years, as brines – solutions of salts in water with reduced melting temperatures – flow up towards the surface, finally ejected in eruptions that form craters on it. And while we do know of at least one salty ocean that supports life, we may have to wait until the 2040s for a potential sample-return mission to Ceres to bring us information on the contents of this subsurface ocean. Besides H2O, asteroids contain many of the minerals and metals that form essential resources for life as we know it, which, given the appropriate technology, are far more easily accessible than those lying below the rocky surfaces of larger moons and planets – but more on that later.

Moving to the Inner Solar System, we find the four planets closest to the Sun: Mercury, Venus, Earth and Mars. These are known as terrestrial planets and have solid surfaces composed primarily of rocks made of silicates (combinations of silicon and oxygen). The similar range of volatiles among four planets at quite different distances from the Sun indicates that there was a great deal of mixing of materials in the turbulent early Inner Solar System – unlike their counterparts in the Outer

Solar System, with which, composition-wise, the Inner planets have less in common.

Mercury is the smallest planet in the Solar System and the closest to the Sun, almost twice as close as Earth. As the fastest, zipping around the Sun at nearly 50 kilometres per second, Mercury is named after the swift-footed Roman messenger god. In the characters used for it in Chinese and Japanese, Mercury is the 'star of water'. And in 2008, with a flyby mission, we did indeed detect the presence of water vapour just above the planet's surface, organic compounds (defined as compounds containing carbon), and water ice inside permanently shadowed craters near its poles. Mercury's axis only tilts slightly, so its polar regions don't receive much direct sunlight. Despite being the closest planet to the Sun, and therefore far more exposed to its heat, could life exist in these relatively mild polar regions?

From the surface of Mercury, the sky would be black due the lack of atmosphere, with the Sun appearing over double the size as from on Earth. A liquid-iron core generates a weak planetary magnetic field, which provides some protection from the charged particles forming solar wind. However, mild is not something this extreme planet does well: Mercury is home to the most dramatic temperature changes in the Solar System. It rotates slowly on its axis over fifty-five days while traversing its eighty-eight-day orbit around the Sun. Surface temperatures exceed 420 degrees Celsius by day while, without a significant atmosphere, at night temperatures plummet to negative 180 degrees, corresponding to a temperature shift of 600 degrees. By comparison, the outside of the International Space Station (ISS) experiences temperatures ranging from 121 to negative 157 degrees Celsius, while on Earth the global range is less than 150 degrees. It remains to be determined whether life can eke out an existence in a potential sweet spot between the bottom of a crater in the cryogenic shade and exposed surface areas subject to periodic scorching. But what of the other planets in the Inner Solar System?

While Mercury may not be the easiest place to live, the next planet from the Sun is no less harsh. Of all the planets, Venus passes closest to Earth, and it is the brightest object in the night sky after our Moon. With some key similarities to Earth – including its composition and distance from the Sun, size, mass and therefore gravity, which is 91 percent of Earth's – at first glance it could look like a good candidate for life beyond Earth.

And until as recently as 700 million years ago, it may well have been just that. There are some indications that Venus had surface liquid water and potentially favourable conditions for life for as long as a couple of billion years in its early history. That is, until things took a turn for the more extreme, making it exceedingly difficult to determine whether or not life could once have thrived there.

As our young Sun grew brighter, the shallow oceans potentially existing on Venus' surface would have begun to evaporate. Water vapour is a greenhouse gas that would have trapped even more heat, resulting in further temperature increases and faster evaporation. Eventually, with no water left on the surface, water molecules in the upper atmosphere would have been broken apart by ultraviolet light, with the hydrogen, a light and thus notoriously flighty molecule, escaping to space as carbon dioxide built up in the atmosphere. All of this, we believe, led to a so-called runaway greenhouse effect and the surface conditions that prevail today. Since the entire planet is enveloped by an opaque layer of highly reflective clouds of sulphuric acid in the upper atmosphere (hence its brightness as seen from Earth), it was only once we had the ability to send probes there that we could discover exactly how extreme things can get on our neighbouring planet.

In 1967 the Venera 5 probe entered the atmosphere of Venus, performing the first in situ analysis of the environment of another planet. In the next few years, further Venera probes captured our only images of Venus' surface, and before ceasing function after about an hour confirmed temperatures of over 450 degrees Celsius: an extremely dense atmosphere of mostly carbon dioxide with a surface pressure equivalent to being almost a kilometre underwater. In 1985, the first successful long-duration flights were achieved on another planet by two instrumented balloons in Venus' atmosphere at an altitude of around 50 kilometres, each travelling almost a third of the way around the planet and collecting data on the conditions there. While we have not landed on

Venus again since 1985, a range of subsequent flyby as well as orbital missions have confirmed in further detail that surface conditions are far from habitable for any life as we know it.

However, the announcement in 2020 of the detection of phosphine in Venus' upper atmosphere has revived discussion around potential habitability above the surface: the gravity, atmospheric protection from radiation, as well as pressure and temperature at just over 50 kilometres above the surface are similar to those on Earth; and furthermore, useful resources like carbon dioxide and nitrogen are present there. Phosphine is a chemical compound made up of phosphorus and hydrogen, and here on Earth it's typically generated by microorganisms living in low-oxygen environments. While the detection of phosphine, and more recently also ammonia, in Venus' atmosphere could be an indication of extraterrestrial life taking refuge from the hellish surface conditions in the clouds, it could also be the result of other processes that are not yet fully understood. Among the many challenges for further investigation or, better yet, the possibility of human presence on Venus at this altitude (although they are not unsurmountable) are the corrosive amounts of sulphuric acid droplets present in the atmosphere. While several orbital missions to Venus are scheduled for the next few years, we may be waiting until the 2030s for the next probe to land there.

Considering the amount of material exchange that continues to take place in the Inner Solar System

– thousands of tonnes of space debris in the form of meteorites fall to Earth each year – it is not impossible that hardy life forms existing on one planet were distributed to others in the vicinity. Both of our neighbours appear to have hosted liquid water oceans for much of their early existence (more on Mars soon); if they also hosted life, we may discover that early microscopic life forms on Earth are in fact part of a lineage that extends over more than one planet. Happily for us, the surface of Mars is a far easier place to investigate than that of Venus, and also contains a much longer record of events, since Venus' surface was covered by new volcanic lava just hundreds of millions years ago, while Mars has extant surface features perhaps even older than 4 billion years, constituting a preserved record from the very early days of the Solar System. Although slightly further than Venus at its closest, it is nevertheless visible with the naked eye as an orange spot in the night sky, a mere few light minutes away from Earth.

Worlds within reach: our celestial neighbours

Our remote investigation of a range of far-flung worlds in our System has shed some light on the kind of neighbourhood in which we find ourselves, but with our limited perspective of reality more questions remain than answers. Out of the weird and wonderful worlds traversing our night sky, given our current capabilities, arguably just three are places we can really get to know:

Mars, the Moon, and of course, although many mysteries prevail yet, our home planet Earth. The Moon and Mars both contain water ice on their surfaces, are rich in information on the history of the Solar System, and may also hold as yet undiscovered clues about our origins.

While we've never been there, we have detailed knowledge of the surface of Mars from decades of data acquired from Earth-based and orbital observations, as well as landers and rovers with direct experience of its surface conditions. Additionally, while we await sample-return missions to Mars in the coming years, hundreds of meteorites retrieved on Earth have been identified as Martian, further contributing to our knowledge of the planet. I even have a piece of one of them (given to me by palaeoanthropologist Francis Thackeray), identifiable by characteristics including the correlation between the air bubbles in the rock and the known atmosphere on Mars.

At on average over a couple of hundred million kilometres, Mars is several hundred times further away than the Moon. In 2019 I travelled quite a bit, almost a million kilometres: equivalent to circumnavigating the globe more than twenty times. So I would need to keep up the pace for another 200 years to rack up the same mileage as a (one-way) Mars trip! Even the rust-coloured light travelling from the surface of the planet takes ten minutes to reach us. Yet in 2020, in spite of the COVID-19-related lockdowns implemented around the world just months before, three missions departed from Earth to Mars, all arriving successfully around seven months later and furthering our knowledge of the planet next door.

Mars is just over half as wide as Earth; its rust colour is due to the iron oxide content in the sand. The thin atmosphere is unbreathable, both because of the pressure, which is less than a hundredth of sea-level pressure on Earth, and its composition, which is 96 percent carbon dioxide, along with small amounts of argon and molecular nitrogen. The gravity is a little over a third of Earth's (or more than double the Moon's). A scoop of the surface sand contains on average a few percent by mass of water ice, as measured by the Curiosity rover, while thick sheets of relatively pure water ice have been detected as little as a few feet down, depending on the location. There may even be liquid water below the surface; recent studies indicate the potential presence of a liquid lake perhaps as much as 20 kilometres across under the south-pole region of Mars. Overall, temperatures on the surface are typically well below zero, with an average of around negative 63 degrees Celsius, owing to the composition of its atmosphere and its distance from the Sun. However, thanks to Mars', albeit thin, atmosphere, conditions there are less extreme than they are on the Moon, with temperatures ranging from negative 150 to 20 degrees Celsius, and, while significantly higher than on Earth, lower radiation levels.

One remarkable similarity to Earth is the day–night cycle on Mars: just thirty-nine minutes longer than our twenty-four-hour cycle (by comparison, a day on Venus is 243 Earth days). Another similarity is the tilt of each planet's spin axis relative to its orbit around the Sun. Earth's tilt is 23 while Mars' is 25 degrees (by comparison, Venus' is 177 degrees: almost upside down), resulting in moderate seasons over the 687-day Martian year, and mitigating temperature extremes, conditions important for known life. Could these remarkable correspondences between the two neighbouring planets seemingly fundamental for life be purely coincidental? In any case, the similarities are promising both in terms of the search for and the potential establishment of terrestrial life on Mars.

Evidence for ancient Martian oceans is more compelling than for Venus; including the preservation of surface structures like canyons, deltas, coastlines and pebbles that on Earth are geologically formed by the presence and movement of large bodies of water. It is estimated that Mars hosted surface liquid water in the early days of the Solar System around 4 billion years ago, and perhaps up until as recently as 2 billion years ago. Some estimate that this ocean may have been around a kilometre deep, covering the entire planet. And with no plate tectonics, much of Mars' crust remains preserved from this era, now potentially as old as 4.5 billion years – older than the oldest parts of Earth's surface, and a record of the very early days of the Solar System. May life have existed here before emerging on Earth? Are our ancestors Martian?

Observations of magnetic rock material on the surface

indicate that Mars once had a global magnetic field. Today, localised crustal magnetic fields still remain, with contributions from an induced magnetic field similar to that of Venus, producing a unique hybrid Martian magnetosphere. However, back when Mars' global magnetic field failed, its once-thicker atmosphere sustaining warmer and wetter conditions was stripped away, since a magnetic field provides protection against cosmic and solar radiation. What happened to Mars' magnetic field? Theories include a gigantic asteroid impact event that melted the crust of a large section of one of the hemispheres, disrupting the flow of liquid metal in Mars' core necessary for the production of a global magnetic field. The Asteroid Belt may contain remnant debris from this impact. Or it could be that Mars is just too small and far from the Sun to have maintained a molten core for long, with gradual cooling resulting in the disruption of the dynamo that sustained its magnetic field. Either way, without a magnetic field to protect it, solar winds stripped away significant amounts of the planet's atmosphere over time, resulting in current surface conditions inhospitable to known life. What about below the surface? While recent periodic detections of methane in the atmosphere of Mars may indicate the presence of some kind of microbial life living beneath the surface, or perhaps be a result of volcanism or hydrothermal activity, subsurface investigations are difficult to perform remotely, and as yet we have no direct indication of life either past or present on Mars. Before diving into more detail on the search for life on Mars, let's first consider the only world we humans have visited, so far, beyond our own planet: the Moon.

Earth's natural satellite, our Moon, orbits at a distance of around 400,000 kilometres, or just over a light second, and has a total area a little bigger than the continent of Africa. Before we set foot on the Moon, we weren't sure exactly what it was made of. And we still aren't sure exactly where it came from. One theory is that a collision of the Earth with another small planet around the size of Mars occurred, and the debris from this impact collected in an orbit around Earth to form the Moon. Local analysis, as well as samples brought back to Earth during the crewed Apollo missions, finally shed some light on its surface composition. Then, since 2007, a series of Chinese missions named after the Chinese goddess of the Moon, Chang'e, have made rapid leaps in lunar exploration. In 2020, the Chang'e-5 mission returned samples to Earth from the nearside of the Moon for the first time since Luna 24 in 1976, while Chang'e-6 brought back the first ever samples from the farside in 2024, revealing different characteristics, including lower density, from nearside samples that may change our ideas about the origins and evolution of our Moon. Further research is currently underway by Chinese teams as well as the international groups (bar the US, due to their own restrictions on bilateral cooperation with China) with whom the samples are already being shared.

Overall, our data and samples show that the Moon's surface composition is not very different to that of Earth; containing largely similar volatiles, minerals and metals, and lending support for the theory that the Moon was once part of Earth. However, the recent detection in samples collected by Chang'e-5 of layered graphene, sheets of single-atom-thick carbon, imply some kind of local carbon capture process, seemingly in opposition to the idea that the Moon was formed by an impact event with Earth. Nonetheless, importantly for upcoming plans for the establishment of permanent lunar research bases, we have detected water ice in the permanently shadowed craters on the Moon's surface, in particular near the south pole, a region promising to be bustling with both robotic and human activity in the coming years. All in all, our investigations have revealed the Moon to be a rather extreme place, with temperatures ranging over hundreds of degrees Celsius during fortnight-long days and nights, from around negative 250 in permanently shadowed locations near the poles to maximums near the equator of up to 120 degrees Celsius during the lunar day; no atmosphere, gravity just a sixth of and surface radiation more than 100 times that of Earth – in fact, the harshness of the environment on the Moon is rivalled by few other places in the Solar System, meaning it's unlikely to host life as we know it. Until 2019, that is.

Humans haven't set foot on the Moon since the last Apollo mission in 1972. Then in 2019, life was indeed detected on the lunar surface, but not for very long. The Moon is tidally locked to Earth, meaning that we only ever see one side of it. Thanks to Pink Floyd for music of otherworldly brilliance, but not for perpetuating the fallacy that the farside of the Moon is dark; it also experiences a day–night cycle of twenty-eight Earth days. Finally in 2019, Chang'e-4 became the first mission to visit the lunar farside, simultaneously demonstrating crop growth there in a small sealed biosphere. The Lunar Micro Ecosystem experiment consisted of an insulated aluminium container 20 centimetres tall, pressurised to Earth sea level and filled with an Earth-like atmosphere, with an internal power supply, irrigation system, seedbed and camera, where cotton, potato, rockcress and rape seeds were placed, along with fly eggs, yeast and 18 millilitres of water. The idea was that the flies would hatch, producing carbon dioxide to feed the plants, whose oxygen emission (stimulated by sunlight entering the payload's interior through a light guide) could then be utilised by the flies. And in January 2019, a cotton seed did indeed germinate: the furthest demonstration of terrestrial plant life from Earth. Unfortunately, the organisms did not make it through the subsequent lunar night, when during systems hibernation all but batteries protecting the electronics were powered down and temperatures plummeted below negative 170 degrees Celsius. All payload functionality was ceased in May 2019.

While a cotton seedling lived and died in an environment more foreign than any plant from Earth had ever experienced before, in fact it may not have been alone. In April 2019, in violation of international planetary protection regulations, a payload on board the private mission Beresheet – which included a cargo of some of the hardiest known creatures – crash-landed into the lunar surface. While the fate of the tardigrades is unknown (we'll return to them and other so-called extremophiles in more detail later), we do know that life on the Moon is tough to sustain. On the other hand, however, without the plate tectonics, water, weather and other dynamics constantly reshaping Earth's surface, the largely preserved surface of the Moon provides a way to study the history of Earth and the life here. For example, meteorites that may have distributed primordial life between planets in the Inner System may have also landed and still be there. What we do know is that the extremity of the Moon provides an ideal training ground just a light second away from Earth for more distant destinations, and a humble cotton seedling is one of the most important milestones yet towards the off-world biospheres necessary to support the planned permanent lunar human presence in coming years.

While our Solar System hosts life in at least one location, its origins and distribution remain a mystery. Water, in ice form or, tantalisingly, in the form of subsurface oceans as well as many of the other resources required for life as we know it abound in our System; however, we have not yet discovered irrefutable evidence of biological activity beyond our own planet, whether ancestral or distinct to terrestrial life. So our journey to the outer reaches of the Universe has brought us back to the one thing we do know about ourselves and our origins: we are from Earth. Let's take a look at the conditions that make it habitable for our and myriad other species, and see what clues we can find on how life may have emerged.

The unique conditions here on Earth are powered by a combination of favourable circumstances: our proximity to the Sun, the presence of our Moon, as well as the size, composition and axial tilt of our planet. These conditions make it possible for large bodies of liquid water to be sustained on the Earth's surface. But where does Earth's water – so precious for all known life – come from? When we look closely into the seawater on our planet and into the nuclei of the hydrogen atoms there, we see that there are more neutrons than we would expect for an ocean sunlit for 4.6 billion years. Hydrogen is the simplest element, with just one proton in its nucleus; a hydrogen atom with an additional neutron in its nucleus is called deuterium. Since energy is required to add this extra neutron, the observed prevalence of deuterium reveals that our water has been exposed to more light than the Sun has yet produced in its entire lifetime. Much of the water on our planet was therefore energised by foreign stars, and delivered by icy comets and asteroids that have impacted our surface. So the water we contain is comprised not only of atoms as old as the Universe, but also molecules older than our Sun!

Alongside the massive quantities of liquid water readily available on its surface, Earth has another unique feature: its magnetic field. This is produced by electric currents in our planet's liquid iron core and stretches far out into space, protecting our atmosphere and surface from radiation from the Sun and from the cosmic rays originating from beyond our Solar System. Recent analysis of rocks as old as 3.7 billion years from Greenland indicate that the strength of this field has remained relatively constant over the lifespan of the planet. In this protected environment, biology has been born and flourished. And although the detection of methane and phosphine in the atmospheres of Mars and Venus, respectively, has triggered a lot of discussion around whether life is involved, we have not (yet) confirmed the presence of life on any other celestial body in our Solar System. On Earth, by comparison, the green jungles on the land easily visible from Earth orbit, as well as vast algal populations in our oceans, play a central role in regulating temperature and rainfall as well as our breathable air, creating a feedback loop where life can thrive. But things have not always been easy for life on this planet. With the grand goal of understanding life, its origins and potential distribution in the Universe, let's go into a bit more detail on what we know about the history of life on Earth.

2

OUR BIOSPHERE

Earth is referred to as the Blue Planet, but an even more striking feature than our oceans are the prolific living networks here, which may have appeared soon after the formation of the planet and are even visible from space. Since we have not yet found life anywhere else, the study of living organisms, for now, is limited to terrestrial life. Looking back in time on Earth may shed light on how life emerged here. Also, understanding the often extreme conditions under which biology evolved and proliferated on our own planet may provide clues as to where best to look for life beyond Earth.

The certainty of motion of the celestial bodies in our Solar System may convey a sense of timelessness; however, things have not always been this way. The Solar System formed around 4.6 billion years ago from the gravitational collapse – potentially due to a nearby supernova event – of a dense interstellar cloud consisting mostly of hydrogen, some helium and small amounts of heavier elements produced by previous generations of stars. At the centre of this now-spinning disc of dust and gas, a hot young star formed, while the gradual accumulation of particles in the remaining cloud gravitated into more massive objects that were spectacularly ejected, destroyed or merged, until the planets and moons with which we are familiar in our System remained.

Devastating impacts with asteroids hundreds of kilometres in diameter coupled with violent volcanic eruptions left little of the original features of the surface of our newly formed planet intact. Earth would have been unrecognisable until a few hundred million years later, when things began to quieten down. As the surface cooled, bodies of liquid water began to collect; here the story of life on Earth begins.

Life started small

As we have seen, Earth is the only place we know of in the Universe to harbour life. Moreover, our planet is teeming with it. The most abundant organisms present on our planet are small; we estimate the existence of a trillion species of microorganisms on Earth, with 99.999 percent of them yet to be discovered. Microorganisms with a size on a scale of millionths of a metre or smaller are called microbes, and proliferate on most parts of the Earth, including inside our bodies – less than half of the cells in our bodies are in fact human! With tantalising evidence that a range of celestial bodies in our Solar System have subsurface oceans, it's exciting to discover that life on Earth may have emerged in the sea.

Our biosphere

Direct evidence of the very first life forms on Earth relies on the difficult task of distinguishing complex geological and chemical structures from simple biological ones, in the oldest preserved places we can find. Earth was formed around 4.5 billion years ago, and tumultuous conditions on our young planet have left little intact for us to study from so long ago. In the first few hundred million years of Earth's life, cataclysmic bombardment by meteorites, as well as erupting volcanoes, resulted in a molten landscape beneath an atmosphere rich in volcanic ejections, likely including hydrogen sulphide, methane and perhaps hundreds of times today's carbon dioxide levels. Life may have emerged and been wiped out several times during this period. As Earth's surface cooled and solidified, bodies of water collected, perhaps as early as 4.4 billion years ago. This is important because a liquid has properties that support the complex chemistry required for life. A gas is usually too dilute and a solid too rigid for a sufficient variety of events to take place; a liquid is the happy medium between the two, and this is where the first simple biological systems probably emerged on Earth. One theory is that hydrothermal vents in Earth's early oceans supported its first biological communities.

Life requires resources, in the form of both energy and matter, and hydrothermal vents may have provided both to the earliest life forms on Earth. Associated with volcanic activity, hydrothermal vents are places where geothermally heated water and dissolved minerals discharge from cracks in the seabed. What appear to be microfossils found in rocks that may have been hydro-thermal vents on an ancient seabed in north-eastern America could be as old as 4.3 billion years, the oldest yet evidence of life on Earth. Researchers continue to search for further evidence to support this claim, sug-gesting the rapid emergence of life after the oceans formed. Meanwhile, on our neighbouring planet Mars, features thought to be ancient hydrothermal vents dated at around 3.7 billion years old – a time when the Red Planet is believed to have had both active volcanoes and surface water – are promising places to look for evidence of microbes that may have been living there around the same time that life was gaining a foothold on Earth. Furthermore, hydrothermal vents appear to be active on celestial bodies including Jupiter's moon Europa and Saturn's moon Enceladus. Given the similar conditions, it's not an unreasonable leap of logic to suppose that hydrothermal vents in environments beyond Earth may have been, or may continue to be, home to extraterres-trial microbial life.

On Earth, the complex microbial communities living near hydrothermal vents are fuelled by both the heat and the minerals and other chemicals dissolved in the vent fluids in a process called chemosynthesis. While chemo-synthesis may have emerged first here on our planet, another ancient living process, photosynthesis, would soon begin to fundamentally change the landscape, shaping the evolution of life for the next few billion years in unprecedented ways.

Less contentious than the claims for evidence of life prior to 4 billion years ago are fossils of microbes found in stromatolites dated at around 3.5 billion years old in Western Australia. Stromatolites are formed of layered rock created by microbial communities living in mat-like structures in the shallow ocean waters prevalent around the planet at that time. Powerful tides, with the Moon several times closer to early Earth, would have played a role in prompting life which was undergoing periodic submergence and exposure to the atmosphere to look up to the energy continuously streaming in from space as a resource. Among the microbes that formed these ancient stromatolites were the first photosynthetic organisms, using sunlight to synthesise organic compounds from atmospheric carbon dioxide and, typically, hydrogen-containing compounds as a source of electrons. Stromatolites grow as each sheet of microbes is covered in sediment, causing the movement of microbes upwards towards the light to form new mats, with the dead layers left behind mixing with clay to form rounded rock structures up to around a metre across.

Photosynthesis is an extremely efficient and highly robust process, making use of some of the most abundant and readily available resources on Earth and spanning most of the history of terrestrial life. Early photosynthetic organisms utilised electrons from resources other than water to create fuel from light and carbon dioxide, but without producing oxygen; so-called anoxygenic photosynthesis. For example, some bacteria use hydrogen sulphide, emitting sulphur as a by-product. Then, in a game-changing event perhaps more than 3 billion years ago, cyanobacteria discovered the potential of splitting water; intriguingly, genetic analysis suggests that this may have happened just once. This new-found and highly effective capability proliferated, resulting in the large-scale emission of oxygen into the atmosphere, which simultaneously triggered the first global extinction event and created the conditions under which complex life emerged and thrived.

While chemosynthetic microbial communities may well be lurking in subsurface bodies of liquid water beyond our own planet, the impact that photosynthesis has had on Earth is not something we have seen elsewhere; this ancient process has dramatically shaped our environment to become a place where abundant life, including us, can thrive. The air we breathe, the food we eat and the fossil fuel we continue to burn are all products of the living systems that use water and carbon dioxide to store sunlight in biomass.

Great change drives complexity

A fundamental characteristic of life is the ability to reproduce and to pass on genetic traits or variations in traits to offspring. When conditions change on scales that are short compared with the generational time-frames of this kind of adaptation, species that are no longer able to procure sufficient resources can become extinct. Sometimes, change is so widespread and sudden that many species become extinct all at once: an extinction event. These events can have numerous causes, but they are always accompanied by a severe and often rapid decrease in biodiversity on a global scale. In fact, extinction is a more common outcome for a species than we may imagine: in the history of life on Earth, which has endured at least half a dozen global extinction events, we estimate that more than 99 percent of all organisms that have ever lived are now extinct.

By examining the fossil record we can identify when such events happened in the past, and sometimes determine what caused the sudden change in the environment. Extinction events have played a fundamental role in shaping the biosphere; a rapid decrease in diversity typically also drives an explosion of complexity, and with whole populations of organisms out of the way, new resource niches open up for new innovations and collaborations in living processes. What was likely the first major extinction event on Earth happened over 2 billion years ago – an epoch when complex life emerged. Perhaps this is more than a coincidence; let's take a closer look.

Up until around 3.5 billion years ago, the atmosphere was almost completely oxygen-free and life on Earth was dominated by single-celled organisms, also called prokaryotes, feeding on things like hydrogen, sulphur, iron and other organic molecules. We may assume that, due to our reliance on oxygen, it is a prerequisite for life; but much of Earth's microbial biosphere still survives to this day on oxygen-free processes. Then cyanobacteria started using water as an electron source and initiated the oxygenation of the oceans and atmosphere through the emission of oxygen as a by-product. By around 2.4 billion years ago, oxygen levels reached a tipping point, resulting in the first known large-scale extinction event on the planet: the Great Oxygenation Event. Oxygen forms a range of highly reactive molecules – kinds of free radicals – that can disrupt living processes and cause cell death, and species without protective mechanisms against this new component of the global atmosphere were vulnerable to annihilation. As oxygen levels rose and displaced greenhouse gases like methane, this also triggered one of the earliest ice ages on Earth.

This is not all that happened around this time, though. Further disruptions to the Earth's atmosphere over the next few hundred million years occurred when a series of large rocks smashed into its surface. The oldest known large-impact crater at around 2.2 billion years old, Yarrabubba in Western Australia, was caused by a meteorite around 7 kilometres across. At that time, the region is believed to have been covered in an ice sheet up to 5 kilometres thick. The asteroid strike would have transformed immense amounts of ice into water vapour – sending perhaps hundreds of billions of tonnes of it into the atmosphere, where it would have served as a greenhouse gas trapping heat. Then 2 and 1.8 billion years ago the Vredefort (more on this event later) and Sudbury impacts, respectively, occurred: the two biggest impacts we know of in the history of Earth, likely caused by large asteroids left over from the formation of the Solar System, over 20 kilometres and as much as 15 kilometres wide respectively. This was a turbulent few hundred million years, but it was also the time when an important leap in complexity happened for life on Earth.

While it's difficult to survey microscopic species' abundances, in particular via microfossils that are billions of years old, a decrease in the size of the biosphere by perhaps more than 80 percent is thought to have resulted by around 2 billion years ago, at the end of the so-called Great Oxygenation Event and in the wake of the severe impact events that occurred during the same period. The presence of oxygen, however, is a double-edged sword: while its high reactivity and cooling effect led to a massive reduction in biodiversity, it triggered the emergence and proliferation of complex multicellular life including fungi, plants, animals and eventually us. An added bonus of the newly oxygen-rich environment was the fact that it enabled the conversion of molecular oxygen into ozone by ultraviolet light in the upper atmosphere, providing a protective layer from the Sun's radiation for life on the surface to flourish. But just how did multicellularity emerge?

Early organisms on Earth consisted of two domains of single-celled life, namely bacteria and archaea. As we have seen in stromatolites, cooperation between early microbes, for example in the formation of mat-like communities, goes back at least 3.5 billion years. However, the emergence of complex life required more than just community living to take place. Given the unpredictable environmental conditions on early Earth, the first single-celled microbes needed to be resilient to maintain themselves. An ingenious piece of natural engineering to solve this problem, and protect the internal living structures and processes of the organism from the outside environment, is a feature as remarkable today as it was then: the cell wall. But this wall also had to be permeable, in order for the resources powering these processes to be exchanged with the environment; thus the cell wall also functions as a membrane.

One explanation of the emergence of multicellularity is that over 2 billion years ago, in what may have been the most important merger of all time (termed the endosymbiotic hypothesis), archaea and bacteria joined forces to give rise to an entirely new building block of life: the eukaryotic cell. According to the theory, this was realised when a bacterium – one that had learned to operate in the new oxygen-rich environment – found itself in a surprising and unknown location: inside the cell wall of an archaeon. Both the bacterium and the archaeon survived this new state of affairs; the engulfed bacterium was not digested, but remained protected from the unpredictable external environment within the host cell while enabling more complex metabolism (much like the plethora of bacteria that live in our stomachs and various other regions of our bodies). These encapsulated bacteria appear to have eventually become the mitochondria that generate energy for the cell, a key characteristic of multicellular life. Similarly, encapsulated cyanobacteria are thought to be the origin of the chloroplasts that perform photosynthesis in the cells of algae and plants, while the nitroplast found in some species of algae may have evolved 100 million years ago when a nitrogen-fixing bacterium was enveloped by an algal cell. The origin of the cell nucleus, another defining feature of the cells that make up multicellular life, remains an open question, but some have proposed that this centre of information, controlling and regulating all activities of the cell, has viral origins; yet another symbiotic collaborator potentially in the picture.

Genetic analysis of complex life suggests, rather compellingly, that the eukaryotic cell evolved in a single event from this complex cell with its contained organelles, including mitochondria, chloroplasts and the cell nucleus; multicellular living systems then emerged in multiple different ways, giving rise to all terrestrial fungi, plants and animals. And so we can see how, in the face of great changes to the environment on Earth, diverse cooperation and joining of forces within Earth's living network gave rise to complex eukaryotic life. However, this is not the end of the story of life on Earth – of course, it is only just the beginning. These early life forms had to survive and overcome many challenges in order for us to find ourselves on the planet billions of years later.

Recent research indicates that land fungi had evolved by at least 1.3 billion years ago, breaking down rocks for mineral resources and gradually turning them into what would become soil for land plants that emerged a few hundred million years later. Then, around 540 million years ago, an explosion of complex multicellular life, including many major plant and animal groups alive today, occurred. Some think that another steep rise in oxygen sparked the change, as animals depend on oxygen which, due to its highly reactive nature, is an energy-rich resource. The process of metabolising food with oxygen releases a lot more energy than most non-oxygenic processes, and animals rely on this potent, controlled combustion to drive energy-intensive innovations such as muscles and nervous systems, as well as the mineralised shells, exoskeletons and teeth necessary for defence and attack in this new era. Others say that the sudden emergence of the ancestors of practically all animals sprang from the development of some key evolutionary innovations, such as vision, enabling the emergence of carnivory. Preying on other organisms is a rich source of nutrients likely to have been a major trigger to the huge diversity of complex life that resulted.

Yet another hypothesis is that the explosion of complexity is correlated with a series of global freezing and warming events that took place around the same time. There is evidence that Earth got really cold for 100 million years, around 700 million years ago – so cold that its entire surface may have been covered with ice, up to a kilometre thick, more than once. Evidence for so-called 'Snowball Earth' includes rock studies in the

Namib Desert of Southern Africa, where rocks showing ancient glacial activity are interspersed with limestone, which typically forms in the warmest parts of the ocean; and also rocks that contain magnetic evidence of having been formed at the equator, while showing signs of having been in the presence of extensive ice cover. Ice reflects sunlight very efficiently back into space, and so once the freeze sets in temperatures continue to plummet.

So how would any life have survived Snowball Earth? When water freezes, due to quantum mechanical effects, it can form a range of different kinds of crystalline forms, including the ice with which we are most familiar and more than twenty others that we know of so far – a new type called 'ice o' was announced in mid-2024. We don't fully understand why the particular crystalline form that floats on liquid water as opposed to sinking – the one we call ice – occurs primarily here on Earth. Luckily it does, however, as this characteristic enables life to prevail in the warmer liquid conditions beneath ice sheets, rather than being frozen to death from the bottom up when surface temperatures plummet.

It is believed that a series of underwater volcanic eruptions finally disrupted Snowball Earth, with this period of freezing paving the way for the burgeoning of complex life from around 540 million years ago. Once again, extreme conditions drove novel adaptations that led to an explosive development in the complexity of life navigating this era. The volcanic activity that ended the deep freeze is thought in turn to have caused a severe extinction event around 510 million years ago, which drove further complexity. A pattern is appearing.

Let's look more closely at how a rapid increase in biological complexity can result from large-scale shifts in the Earth's climate. Experiments show that in a test tube, single-celled life placed under conditions of resource constraints can begin to evolve multicellularity in surprisingly short timeframes. For single-celled organisms, the preservation of the individual cell allows for genes to be passed on to the next generation. Multicellularity, on the other hand, requires the de-prioritisation of the self-interest of each cell, in favour of the survival of the group of cells making up the organism. Fungi are rather anomalous life forms; among other things, some of which we'll get to later, there are members of the fungal kingdom that are eukaryotic but also single-celled, making interesting case studies for the emergence of intercellular collaboration. Yeasts – the same ones we've been using in bakeries and breweries for thousands of years – are an example of such unicellular fungi. In extreme conditions, meaning that access to resources is restricted, a few dozen generations are all that is required for single-celled yeasts to evolve into larger co-dependent clusters. The daughter clusters then only create their own offspring once they reach a similar-size cluster as their parents, an indication that each cluster begins to behave as an individual organism. Evidence of division of labour, an essential characteristic for more complex multicellular life forms, can be seen in the emergence of cells specialised to die in order for the yeast structures to reproduce. We see here in laboratory conditions how extreme environments and the resulting resource scarcity drive increased collaboration and complexification in living networks.

What may have been the largest extinction in Earth's history was not, as many might believe, the one that exterminated the dinosaurs. Instead, it occurred some 252 million years ago and is referred to today, charmingly, as the 'Great Dying'. This happened before the emergence of dinosaurs, wiping out nearly all marine species and many plants and animals, and is believed to have been caused by a series of massive volcanic eruptions in Siberia. The continuous explosion of large amounts of greenhouse gases like carbon dioxide into the atmosphere would have suddenly elevated global temperatures once more, and when carbon dioxide dissolves in the ocean, high acidity levels and a lack of oxygen in the water result. This was such a rapid environmental change, in particular in the oceans, that as much as 96 percent of marine species around at the time did not survive.

The Great Dying was in turn followed by what is perhaps the best-known extinction event: the one that annihilated the dinosaurs and many other species with it around 66 million years ago. We have convincing evidence of what likely happened many times in Earth's history: a major impact event disrupted the climate, resulting in the extinction of large numbers of species not able to adapt to the new conditions that resulted. Creatures at the top of the food chain are most susceptible to such events: they are typically more complex and evolved for specific conditions; also larger animals are at increased risk as they take longer to reach maturity and reproduce than smaller organisms, meaning genetic adaptation to change happens over longer timeframes.

Our evidence for this event includes an unusually high amount of iridium capping rock layers containing dinosaur fossils. Iridium is relatively rare in Earth's crust but is more abundant in stony meteorites, which leads us to conclude that the mass-extinction was caused by an extraterrestrial object. Furthermore, an impact crater found in Central America dated at 66 million years old indicates that an object at least 11 kilometres wide hit the Earth, which would have filled the atmosphere with gas, dust and debris and drastically altered the climate, blocking out sunlight for as long as eighteen months and cooling the planet for the next decade.

Suffice to say, life on Earth has been through the wringer; but, while not without drama, we have seen that, in some shape or form at least, life has been around for most of the existence of our planet in spite of, or even fuelled by, the often extreme conditions. When we look beyond Earth, conditions are often even more extreme. What does this mean in the quest to understand our origins? While we must admit that our knowledge of even our own Solar System is limited, increased sophistication in our search for life beyond Earth has not revealed evidence of the impact visible from space that photosynthesis has had on our surface and atmosphere, and indeed on all subsequently evolving life. Perhaps our search has been too superficial. Subsurface oceans on celestial bodies in our own System may well yet be shown to host biological communities, or even extreme kinds of microbes familiar to us. Such a discovery would forever change our thinking about life and its abundance in the Universe.

However, in our pursuit of the question of where we come from – in other words, in our quest to understand how life originated – the phenomenological study of the evolution of life on Earth from small microbes to more complex animals is hampered by several critical developments appearing to have happened just once. That the eukaryotic cell, the water-splitting cyanobacterium and indeed the first living organism may each have manifested in singular events provides us with minimal information indeed to come up with a general theory of life and its origins on Earth.

Our universal common ancestor

Other than the common requirement for energy and nutrients, in particular liquid water, underlying the remarkable continuity of life is a molecule that has consistently connected all living forms on Earth: deoxyribonucleic acid, or DNA. While we have not yet solved the puzzle of precisely when and where it originated, never mind how, one of the most beautiful and impressive discoveries that contributed to our understanding of living things is that we all, from microbes to humanity, contain the same information-carrying structure. Our study of this molecule, which contains the genetic instructions for all known organisms, has also shed some light on our quest to understand our origins.

Resolving the structure of DNA in the 1950s was a major breakthrough in the field of genetics. As chemist Rosalind Franklin said, 'the helical structure has novel features of considerable biologic interest'. 'Considerable' may have been the understatement of the century. Through her work using X-rays she established that the DNA molecule exists in a helical conformation, laying the foundation for biologists James Watson and Francis Crick's conclusion in 1953 that DNA has a double-helix structure. This structure makes DNA very stable: a useful property for the molecule ensuring the persistence of life. The two chains that coil round each other to form this double helix carry genetic instructions for the development, functioning and growth of all known living things. Using sequences of just four chemical bases – adenine, guanine, cytosine and thymine – DNA controls the combination of the right building blocks in the right order to build things like the proteins making up all known life.

Since the vast majority of each species' DNA is the same in all individual organisms, it is possible to determine the code for the entire sequence, or genome, for that species. Starting with the sequencing of the genomes of simple bacteria and archaea, by the year 2000 we had successfully determined the majority of the 3.2 billion base pairs making up the human genome: a total of nearly 800 megabytes of information. The celebration that we had discovered the instruction manual for life was, however, rapidly tempered by research showing how much we still have to learn about genetics. A range of results in a field called epigenetics show that the way the information contained in DNA is read by a living organism depends intrinsically on the external as well as the internal environment of the organism. Behaviour, diet and exposure to pollutants and toxins have been shown to be associated with a modification of the genetic activity impacting health and longevity, while studies of advanced meditators measure different patterns of DNA activation during periods of meditation compared with normal mind states. DNA is more than just an immutable set of instructions: myriad complex environmental factors can influence which genes are turned off or on during the lifetime of an organism. Our understanding even of the sequence itself is incomplete, shown every time that examples of previously designated 'junk' DNA are in fact found to be responsible for critical living processes. For example, ancient invasive viral genetic elements called transposons – which constitute nearly half of mammalian DNA – were previously thought to have no biological function. Recent experiments have shown, however, that transposons play a critical role in embryonic development in mice and perhaps all mammals, to the extent that the removal of a particular transposon resulted in the death of half of the unborn baby mice studied.

Can our incomplete but growing understanding of genetics shed any light on the origins of life? From sequencing different species' DNA, as well as estimating the rate of the random mutations that contribute to the genetic divergence between species, we are able to classify organisms based on evolutionary relationships. Closely related organisms in this tree of life will have highly similar genetic sequences, while distantly related organisms will have sequences that have diverged because of the accumulated changes since their evolution from a common ancestor. Along with divergence, genetic analysis has also revealed shared genetic segments between organisms as different as us and bacteria; revealing that life on Earth is one big terrestrial family.

Going back in time, this has allowed us to construct a chronological tree, suggesting that all life on Earth is related and derives from a single last universal common ancestor, or LUCA, that existed around 4 billion years ago. More recent analysis puts this date back even further to 4.2 billion years. This is in remarkable correlation with evidence from the fossil record of when life emerged on Earth. Evolutionary biologist Charles Darwin once declared that 'probably all the organic beings which have ever lived on this earth have descended from some one primordial form, into which life was first breathed'. If terrestrial biology did indeed spontaneously emerge from complex chemical systems, we might suppose that this event would have taken place more than once, rather than the seemingly precarious scenario where only one such organism gave birth to all life on Earth.

One compelling explanation for a single common ancestor is the theory of panspermia, which proposes that the components of life, and moreover perhaps microbial life itself, are continuously distributed to planets by comets and meteorites. Could one such space rock have housed LUCA, and in fact have been the original source of life to have arrived and multiplied on Earth? This occurrence, however, does not answer our question of how life emerged in the first place; it just pushes back the location to somewhere beyond Earth, perhaps to a subsurface ocean somewhere right here in our System. But before going off-world, let's delve even deeper into terrestrial life to try to understand where it came from, into the atoms and molecules that make up living systems.

What is life? Zooming in

Human imagination knows no bounds. Neither does reality, it would seem. Unprecedented feats of engineering in space exploration have enabled us to observe and investigate our position in the Universe further than ever before, as well as to describe the behaviour of the smallest known constituents of reality in the most accurate scientific theory to date, quantum physics. In spite of this grand understanding of reality on mind-bendingly vast scales, from trillionths of a foot to trillions of steps, there is a phenomenon that lies somewhere in the middle that eludes scientific understanding to date: life.

After years of studying theoretical quantum physics, I was used to the notion of the existence of things I have never seen and can never see. When a guest lecturer came to our campus to give a talk on quantum effects in photosynthesis, I was sold. Finally, I thought, when people ask me what I am studying, I can point out of the window. Given that we have a mainstream theory of the origins of the Universe, when I decided to do my PhD in quantum biology I was thinking about, in my opinion, the biggest open question in science: how did life originate? From a theoretical perspective, defining life is fundamental to describing how it emerges. Thankfully, Francesco Petruccione, leading world expert in open quantum systems and my PhD supervisor at the time, always encourages new ideas and supported my move from studying quantum cryptography to studying, up close, one of life's earliest, most beautiful and most prolific processes: photosynthesis.

What is life? I asked. And as living systems ourselves, can we ever really know? Definitions of life often include lists of observed capabilities, like metabolism, growth, reaction to stimuli, reproduction and so on. But perhaps the most satisfying definition of life we have arrived at is: something that is not dead. As living organisms, we have the instinctive ability to distinguish between living and dead things, a useful skill on a planet where life is abundant. The lack of clarity on what exactly life is means controversy for the status of things like viruses and transposons, and soon artificial intelligence, as well as strategies for searching for and potentially confirming the presence of life beyond Earth.

In order to arrive at what would qualify as a theory of life, we will need to derive a clear understanding of biology in terms of physics and chemistry; we will need to understand how living things emerge and differ from the matter of which they are made.

Have impressive developments in the biological sciences brought us any closer to understanding life? Can we pinpoint what distinguishes an inanimate bunch of molecules from the collection of molecules that make up a living organism? To understand the properties of molecules and how they interact, we will need to consider them through the lens of quantum mechanics. Quantum mechanics describes, often counter-intuitively, phenomena on very small scales, millions of times smaller than what we can see with the naked eye: objects like molecules, which are made up of atoms, as well as fundamental particles with no regular size at all like electrons and photons. The movement and behaviour of visible objects on the scales we are used to as humans can mostly be sufficiently explained by classical physics; including, for example, the movement of the celestial bodies in our Solar System or, for one of the founders of classical physics, Isaac Newton, the apple falling from the tree. Just as we need new physics, namely general relativity, to explain the phenomena we observe in the Universe on much larger scales, as we zoom into reality quantum physics provides us with a new set of rules for what we see there. Life is fundamentally made up of tiny quantum objects, and so the question is: do we need to delve all the way down into the quantum world to understand exactly what is going on in biology?

Our experience of reality is both limited and generated by our senses. Those with good eyesight can see debris and dust of a minimum of around 25 micrometres, or microns, in size, or less than half the width of an average human hair. This hasn't stopped us wondering what is going on at smaller scales, of course, and although it's only relatively recently that we have been able to delve deeper, philosophers have mused for thousands of years on whether there may be indivisible building blocks from which matter is made; the word 'atom' derives from a Greek term meaning indivisible.

As in probing faraway places with telescopes, glass and optics have also played a central role in our peering down at the very small. With the advent of microscopes in the early 1600s, our investigation of the small was extended beyond the visible world. Using a set of convex lenses to bend light reflected from an object, magnifications of up to a few hundred times enabled us to see the structure of microscopic biological systems like red blood cells, spermatozoa and bacteria, which are just a few microns wide. The resolution of a microscope using a light beam to illuminate a sample is limited too, though this time by wavelength; for visible light this is around a fraction of a micron, or two orders of magnitude more than the resolution of the naked eye.

To 'see' on even smaller scales, a beam of something with shorter wavelengths than visible light is required – something like an electron, for example. Electron microscopes were developed in the 1930s, making use of the relatively new (at that time) discovery that quantum objects like electrons can behave as particles and as waves. By using an electron beam to illuminate a specimen, due to its shorter wavelengths images with magnifications of millions of times – thousands of times the resolution of the eye – can be generated. Unlike massless photons, electrons do have mass, which means shorter wavelengths, but also that they can easily be deflected by comparatively massive gas molecules, and so the sample needs to be placed in a vacuum. Just like human sight, and just like optical microscopes, electron microscopes also have their limitations. To go even deeper, down to the quantum realm on the scale of individual atoms, let's look more closely at what quantum mechanics tells us.

Quantum objects have uniquely quantum properties like spin and vibration; they can behave like particles or waves depending on the context; exist in more than one state at once; share correlations no matter how far apart they are – none of this is observed in the macroscopic world with which we are familiar. For example, in the framework of classical physics an object can't spontaneously pass through a barrier without the energy to overcome it. In the quantum world, though, a particle like an electron can also exist in a wavelike state, enabling just that. The ability of an electron to spontaneously 'tunnel' through something in fact enables us to see into reality on even smaller scales; by scanning the surface of a specimen with a sharp conductive probe, when the distance between them is small enough, the probability of electrons jumping the gap increases, enabling an atomic-level imaging of the surface by monitoring the movement of electrons.

In 1981, the first scanning tunnelling microscope was demonstrated. It used the principle of quantum tunnelling to enable the visualisation and manipulation of a wide range of inorganic and biological samples, from crystals to cells, in environments not strictly requiring vacuum and on the unprecedented scale of individual atoms, corresponding to magnifications of tens of times more than the electron microscope. Using scanning tunnelling microscope technology we have been able to identify, control and move individual atoms, although not terribly quickly. In 1989, the ultra-sharp tip of a scanning tunnelling microscope was used to pick up thirty-five xenon atoms and spell 'IBM' in 5-nanometre-tall letters. It took twenty-two hours.

*

But what has peering down to the scales where atoms and molecules exist taught us about life? Life operates at the interface between the macroscopic reality we are used to and the quantum regime, providing an opportunity to explore the mysterious transition between these worlds. As far back as the early 1900s, when quantum theory was being developed, there was speculation by some of the founders as to whether quantum phenomena could possibly play a role in living systems. If so, it was thought, perhaps quantum physics could provide us with the tools to understand life itself.

In 1932, ten years after Niels Bohr, one of the founders of quantum physics, was awarded the Nobel Prize in Physics for his work on atomic structure, he gave a lecture entitled 'Light and Life', raising the question of whether quantum theory could contribute to a scientific understanding of living things. In attendance was an intrigued Max Delbrück, a young physicist who subsequently contributed to the establishment of the field of molecular biology and won a Nobel Prize in Physiology or Medicine in 1969 for his discoveries in genetics. In his 1944 book *What Is Life?* Erwin Schrödinger, another of the founders of quantum theory, questioned how the physical events taking place within complex living organisms could be accounted for by physics and chemistry, proposing that some kind of aperiodic crystal encodes information that somehow guides the development of the organism. That 'crystal' turned out to be DNA.

Einstein commented that 'one can best feel in dealing

with living things how primitive physics still is'. As we peer deeper into the functioning of living things – equipped with technology built on the basis of our growing understanding of the quantum world – into the atoms and molecules that make life possible, we are finding an increasing number of processes where we need quantum physics to explain what is going on. Quantum biology is the field that deals with such cases. Have developments in physics and in particular quantum biology over the past century brought us any closer to a theory of life?

The vast scale of separation between processes described by quantum mechanics and those typically studied in biological systems, as well as the seemingly different properties of things dead and alive, has meant a traditional distinction between the two bodies of knowledge. During my PhD, I took a rather long walk from the physics to the biology department (typically these buildings are spaced far apart), to chat to people there working on photosynthesis. When I asked what timescales they are looking at, one researcher explained that he monitors marine algal populations over years. I returned, rather disappointed, to my desk and to my study of photosynthesis, which went down to picosecond (a trillionth of a second) timescales ... Only nineteen orders of magnitude separated our studies!

More recently, however, the regimes of applicability of these two areas of science have begun to overlap:

quantum physicists are able to work with systems of increasing complexity, like running experiments on viruses and bacteria, while life scientists can give increasingly detailed explanations of macroscopic phenomena in terms of molecular structures and processes, such as in the field of genetics.

Living organisms, even the most basic ones, are highly complex open systems, which means that things are constantly coming in and going out, making it more complicated to keep track of what you are studying. In living systems, every moment, there are a plethora of interactions going on internally, where a host of nano-machines go about the business of maintaining a state of living within each cell, as well as between organisms and their environments, requiring a continuous exchange of energy and matter to fuel these internal processes.

For much of our (incomplete) understanding of biology, classical physics and chemistry have provided a satisfactory description of what is happening. For example, every schoolchild learns the chemistry of photosynthesis: that carbon dioxide plus water gives sugar plus oxygen in the presence of sunlight. However, when we look deeper, or when we try to build systems that imitate Nature, we realise that there's a lot more going on than meets the eye. And so we use our technology to zoom in to observe things in more detail. We find that there are a range of living processes where quantum mechanics is necessary to accurately describe the full behaviour of the system, photosynthesis being such a case. While these quantum effects may appear hidden, quantum biology addresses cases where their role impacts the overall survival of the organism.

Let's take a look at one of the first living processes to emerge on Earth, in one of the simplest living systems that we know of – bacterial photosynthesis – to see if a deeper understanding of one of the most primitive life forms on Earth can shed light on what life is, and how it emerged on our planet.

A prolific class of organisms, including bacteria, algae and plants, have been using sunlight and abundant terrestrial resources like carbon dioxide and water to produce energy-rich compounds to power cellular activities for as much as 4 billion years. As one of the oldest known living processes, its study is the most well-established area of quantum biology. The idea that quantum effects may play a role in photosynthesis has been around for over a century, but, bar a few early experiments, it is only since the turn of the century that developments in experimental techniques have allowed us to observe the extremely rapid processes involved. And we have seen things only possible in the quantum world playing a role in bacteria and algae, as well as plants. While these quantum effects may appear subtle, they contribute fundamentally to the robustness and efficiency of the process, ensuring the survival of photosynthetic organisms for most of the duration of life on Earth. We'll focus here on bacterial photosynthesis for illustration.

With ultra-fast spectroscopy – basically, taking snapshots of processes completed on extremely fast timescales like picoseconds – we see that, all within a fraction of a millisecond, pigment molecules in the photosynthetic apparatus absorb a photon of sunlight, convert this energy into the movement of an electron from one place to another, and generate a charge separation between the positive 'hole' that is left behind and the negative charge of the electron at its new position. Once the electron and hole are separated by enough distance, a few nanometres, this charge-separated state then acts like a battery, ultimately driving fuel production, including molecules like glucose and adenosine triphosphate (ATP). This light conversion process happens with a quantum efficiency approaching 100 percent, meaning that for every photon absorbed, an electron is transferred; by comparison, our best solar technologies can reach quantum efficiencies of just 50 percent under laboratory conditions. We understand that such high efficiencies are possible in photosynthetic organisms because of intricate quantum interactions between the photons, the electrons, the pigment molecules like chlorophyll where they are absorbed, and the protein environment in which all of this takes place – the proteins in the photosynthetic system vibrate in very specific ways to guide first the light energy and then the electron to exactly where it needs to go.

Using resources so very efficiently is one way to prevail on a planet for 4 billion years. Another useful tactic is to avoid premature death. Our research of the photosynthetic apparatus of anoxygenic purple bacteria revealed a quantum spin-based protection mechanism against oxygen free radicals – the kind of defence mechanism that primitive anoxygenic microbes would have needed to survive the Great Oxygenation Event. Astoundingly, the origin of this protection mechanism is a strategically positioned iron ion; just a single atom of iron missing two electrons has a high enough quantum spin to significantly reduce the chances of cellular death, which for single-celled life forms is indeed death, by oxidation. Also found in a strategic position in the photosynthetic apparatus are two single water molecules, whose spins are believed to play a role in addition to the protein environment in regulating the light-induced electron transport. In spite of our increasingly detailed knowledge of photosynthesis, many open questions about the process remain, particularly in understanding the structure and function of the immediate environment within the organism where this all takes place.

Why go into such detail on bacterial photosynthesis? As we look at life on increasingly small scales, we see that the sophistication of the biological world is a few steps ahead of our own technological capabilities. Unsurprising perhaps, given that life has enjoyed 4 billion years of research and development here on Earth. Even the simplest living systems are mind-bogglingly complex. Firstly, what we see here is that uniquely quantum characteristics, like the quantum vibration of the protein environment and potentially also the quantum spin of the iron ion and the water molecules, play a role in efficiency and defence which are matters of life and death for the bacterium. Photosynthetic single-celled microbes are some of the most primitive living systems, and are well studied. Our detection of quantum effects essential to the functioning of these relatively simple living systems gives a good indication that quantum mechanics are likely involved in a range of processes in the more complex life that evolved from these early organisms. Secondly, the observation that single photons, single electrons, as well as a specific iron ion and a couple of water molecules all appear to play fundamental roles in photosynthesis gives good insight into why modelling living systems is tough: the exceedingly vast numbers of tiny particles directly involved in living processes are impossible to model exactly with our current computing capabilities, and even modelling them approximately can be a bit of a headache.

Modelling exactly the approximately 100 billion atoms in just one simple bacterial cell – that's the same number of stars in the Milky Way – is a feat currently beyond our computational power. A molecular system almost 100 times smaller containing 1.6 billion atoms was modelled on the world's most powerful supercomputer until 2022, the Fugaku, with standard approximations that leave some things out: one second of activity would take more than 300,000 years to simulate. Happily, the researchers are also interested in the dynamics happening on much shorter timeframes, like nanoseconds (billionths of a second), which can be computed in hours.

Not only life, but reality is exceedingly complex when we look at it on a quantum level. Water is the basis of all life on Earth; all cells contain around 70 percent of the stuff. Its structure is simple: two hydrogen atoms bound to one oxygen atom. Yet its bulk behaviour is unique among liquids, and we still do not fully understand how its distinctive properties arise. For example, only recently have we discovered that there are in fact two distinct types of water in the average glass – the quantum spin of the protons in the hydrogen nuclei in each molecule can either be aligned or opposite – and that these configurations have very different properties. The switching between spin states in the two water molecules in the bacterial photosynthetic apparatus just mentioned may act as a gate for electron transport. The quantum mechanical characteristics of water are known to play a role in a range of processes in living cells: for example, quantum mechanical protein–water interactions affect the structure and dynamics of proteins in a water environment in the cell, and likely also play a role in the functioning of the brain, which is 75 percent water. Practitioner of alternative medicine Masaru Emoto claims to have demonstrated that water is shaped by the environment, including human thoughts and emotions; that positive thoughts and emotions produce beautiful symmetrical crystals when water in the vicinity of a human experiencing these emotions is frozen, whereas negative ones don't. While current science cannot explain how human consciousness could interact with the molecule so fundamental for terrestrial life, models of the brain involving quantum spin are an active field of research.

Trying to understand how life emerged on Earth, how collections of atoms and molecules became organisms, in what may have been a singular event, is a difficult problem and not something current science has been able to explain. And perhaps it can't. One thing we do know is that all life on Earth is interrelated, with a common 4-billion-year history detailed in the complexity, and commonality, of each life form's genome. Which means that in trying to understand what life is by studying life on Earth, we are essentially studying a single system, terrestrial life, while trying to derive a general theory from it. It's clear what's missing: a second data point.

3

ARE WE ALONE?

Is there life beyond Earth? Would we recognise it? These are difficult questions to answer theoretically. Because thus far we have only one example of life's emergence, and it's rather difficult to come up with a general theory based on a single data point. We are also limited by a lack of precise knowledge of the conditions under which this happened, in what could be a singular event. The discovery of evidence of life beyond Earth, whether related or completely distinct from life here, would provide that elusive second data point; a giant leap towards understanding ourselves and our place in the Universe. While we haven't yet confirmed the existence of extraterrestrial life, past or present, from our vantage point here on Earth, we have made some interesting progress.

The building blocks of life in space

While there are many weird and wonderful places in our Solar System, the similarity between the conditions here on Earth under which life is thought to have emerged

– in particular, the presence of liquid water coupled with geological activity – and the conditions both on ancient Mars and in current subsurface environments on a range of worlds in our System is a compelling reason to search for simple living systems resembling terrestrial microbial life beyond Earth. From our current position, however, even here in our own System, rocky terrain, thick ice sheets or opaque atmospheres can obscure our view of whether any of these celestial bodies indeed host living organisms. While we await future missions to visit these worlds to send us new information, can our investigations of life on molecular scales play a role?

Panspermia is the theory that the components of life or life itself are continuously delivered to larger celestial bodies by comets and meteorites. Peering down at life on tiny scales has revealed that all known terrestrial life is made up of proteins formed by chains of just twenty different amino acids. Just as the English alphabet contains twenty-six letters, and all words, sentences, paragraphs and books are made up of chains of these letters, these twenty amino acids combine in different ways as the essential components of life on Earth, all of which is encoded in DNA. So, both amino acids and the genetic bases that make up DNA are considered to be important building blocks of life. Molecules that are prerequisites for the formation of these building blocks, so-called prebiotic molecules, are also of interest. A first step towards evidence for panspermia would consist of the detection of some of these building blocks necessary for life in space. While we haven't yet spotted life itself, we have made some interesting discoveries.

Improved observation techniques and new instruments like the James Webb Telescope for scanning the skies beyond our Solar System, also the interstellar space between stars, have revealed a range of molecules known to be prerequisites for life. We have also found evidence closer to home of the building blocks of life forming beyond Earth. The presence of these ingredients in space may give us clues as to how life emerged on Earth, and whether it may have done so elsewhere.

The chemical components required for life, namely the different molecules that make up living organisms, have been around for quite some time. Hydrogen is by far the most abundant element, and was produced along with helium – which is inert, or stable, and doesn't bond easily with other elements – in the first few hundred thousand years after the Universe formed. Oxygen, the third most abundant element, required stars for its production, and the first H_2O molecules were likely established in pockets across the Universe around a billion years after the Big Bang.

Many of the molecules we observe in space are detected in interstellar clouds – gas clouds in between stars in our Galaxy – and are mainly composed of the cosmically abundant elements hydrogen, oxygen, carbon and nitrogen. Carbon is a dominant constituent of the interstellar molecules detected in the Milky Way; in fact, all the interstellar molecules we have detected so far with six or more atoms are carbon-containing, or organic. Since the detection decades ago of simpler molecules (where each element is represented by its chemical symbol) like CO, carbon monoxide; HCN, hydrogen cyanide; NH_3, ammonia; and H_2O, water, the signatures of hundreds of complex organic molecules have been observed, mostly in the radio frequency band in interstellar clouds across most of our Galaxy. More recent detections of prebiotic interstellar molecules include the first sugar, glycolaldehyde, an eight-atom molecule containing hydrogen, oxygen and carbon; the hydrogen cyanide dimer, consisting of two HCN molecules which is a precursor of the genetic base adenine; a precursor to the simplest amino acid, glycine, as well as (tentatively, as described by the authors) glycine itself and later another amino acid, tryptophan. However, controversy around the last two detections suggests that the search for the first amino acid in interstellar space is not over yet.

Closer to home, some of these precursors have even been detected on comets. These small icy rocks orbiting the Sun are thought to originate in the Outer Solar System. When passing close to the Sun, they warm and begin to release their volatiles in a process called outgassing, which creates the characteristic halo, also known as a coma, around the comet's body. In the coma samples collected by the Stardust comet flyby mission of 2004, glycine was found together with other complex prebiotic molecules. The Rosetta mission, launched the same year, also detected glycine, as well as phosphorus, a vital element in cell membranes and DNA, in the coma of the comet on which it performed the first ever comet landing. Then, in 2022, more than ten amino acids were reported to have been detected in the 5.4 grams of sample collected from the Ryugu asteroid 300 million kilometres away and returned to Earth by the Hayabusa2 mission. And analysis of the 200 grams of sample returned to Earth in 2023 from the asteroid Bennu after a seven-year-long voyage on the OSIRIS-REx is still underway, with suggestions that, because of the significant phosphate content, Bennu may be a fragment of an ancient ocean world. While these detections don't tell us immediately how life formed or whether we are alone in the Universe, they are exciting for a number of reasons.

The first sugar detected in space is also a precursor molecule for the formation of RNA. RNA, ribonucleic acid, is present in all known life, working together with DNA to synthesise proteins. The 'RNA world' theory of the origins of life on Earth proposes that RNA came first, before the evolution of DNA and proteins; finding this sugar in space could be evidence for the theory. More generally, the discovery of amino acids both in comets and in an asteroid in our Solar System, as well as in much more remote regions in our Galaxy in interstellar space, is evidence for the idea that the building blocks of life beyond Earth are widespread, giving support to the argument that life in the Universe may be more common than we think. These molecular precursors to life could have been delivered to Earth's early surface water by meteorites, and formed ingredients in the so-called primordial soup from where the first living systems may have emerged.

Furthermore, many meteorites found on Earth (which are mostly remnants of asteroids from just beyond Mars) also contain a range of building blocks of life. For example, over seventy different amino acids (that's an additional fifty or so beyond the amino acids that make up terrestrial life) have been detected in the Murchison meteorite that fell in Australia in 1969. Other well-known examples include the Orgueil and Tagish Lake meteorites, which have also been found to contain amino acids as well as a number of other pre-biotic molecules.

This is great news; we see a whole range of molecules important for life not only in our Solar System but also in interstellar space in our Galaxy. How hard can it be to figure out how they combined into life here, and therefore could have done so elsewhere too?

In 1953, chemists Stanley Miller and Harold Urey recreated what they believed to be representative of an early Earth atmosphere in the laboratory: a gaseous mixture of methane, ammonia, water and hydrogen. To simulate the frequent lightning storms thought to be characteristic of early conditions on the planet, they passed electricity through this test-tube atmosphere. Miller said of the outcome: 'Just turning on the spark in a basic pre-biotic experiment will yield eleven out of twenty amino acids.' The intriguing Miller–Urey experiment produced a wide range of amino acids, and it did so with remarkable ease. Were we on the brink of taking the first steps towards synthesising life in a laboratory?

Further experiments have found that amino acids form in a diverse range of conditions, when ultraviolet light or X-rays are passed through this same gaseous mixture at room temperature, and also when ultraviolet light is shone on the surfaces of small dust grains produced in the laboratory to resemble the icy conditions in interstellar clouds. It is now believed that Earth's early atmosphere was in fact mainly carbon dioxide and water vapour with only small amounts of ammonia and methane, conditions which, in repeats of the Miller–Urey experiment, lead to the formation of trace amounts of amino acids. Nonetheless, these experiments show us that the emergence of the building blocks of life happens spontaneously when energy – in forms from electricity to light – is added in a range of environments as different as cryogenic dust grains to warm organics-rich gases.

The identification of functional quantum effects in some of the oldest and simplest living systems we know of, namely photosynthetic bacteria, suggests that quantum effects also played a role in the emergence of the very first living systems as well as their precursors, and that techniques from quantum biology could be useful for origins-of-life studies. Energy is required for anything interesting to happen; and there are compelling parallels between the analysis of how a range of light frequencies, or even electrical lightning, result in the spontaneous formation of the building blocks of life in both gases and on icy surfaces. Light-induced processes are central to photosynthesis, a deeper understanding of which will likely hold clues about life's blueprint, as well as how life emerges from the complex but inanimate matter of which it is formed.

However, as we've seen, modelling complex systems, especially biological ones, on a quantum level is tough. In fact, our current computational systems are even insufficient to account for all the variety and richness of chemistry (generally simpler than biology) occurring in space. But let's start small. Molecular hydrogen, or H_2, is the smallest and by far the most abundant molecule in the Universe, making up 99.99 percent of molecules in space. Bombarded by high-energy particles from nearby stars, more complex molecules, for example water, then emerge from this basic component. However, since the first detection of H_2 in the interstellar medium in 1970, we have noticed a disparity between the amount of H_2 we see in space and our calculations of how much there should be, based on how we believe it to form: there is more H_2 around than we can explain.

It was suggested that the surfaces of icy dust grains, rather than the largely empty vacuum of space, could serve as catalysts for the formation of molecular hydrogen; hydrogen atoms stuck to the surfaces of these grains would have more likelihood of combining. In the interim, both theory and experiment have revealed that many complex molecules, not only H_2, form primarily on the surface of dust grains in processes where quantum effects, including quantum tunnelling, play an important role. The compositions of actual dust grains – mostly smaller than a micron – are hugely variable, each kind of composition having unique quantum characteristics like vibrations and spin, all adding to the complexity of the process.

One day, during my postdoctoral studies in a field I'd like to call quantum astrobiology, I sat staring at my computer screen taking all of this in. A theoretical understanding of the emergence of life requires a full understanding of the formation of the simplest molecules. If we are to describe scientifically – by that I mean in terms of physics and chemistry – the emergence of biology, we would need to start with at least a comprehensive understanding of chemistry. But if I engaged in the research programme I had in mind – developing an open quantum systems formalism in different environments to account for the formation of molecules from H_2, to H_2O, to prebiotic molecules, to amino acids and genetic bases, to the formation of proteins and DNA – then, with even H_2 formation not solved in the detail I was hoping for, I likely wouldn't still be around for the grand culmination of the development of a theory of the emergence of life from the matter distributed through the Universe. I realised that we don't know much. Or, more positively, as my former PhD supervisor Francesco still says to me, 'We are always learning new things.' Including just how resilient the mysterious phenomenon of life can be in harsh conditions.

Extremophiles: clues to life beyond Earth

The detection of the building blocks of life in a variety of places, from interstellar clouds to meteorites, suggests that life could be far more widespread than we think; these relatively abundant ingredients just need to find a suitable place to combine. But the panspermia hypothesis also goes one step further: proposing that life itself is distributed through the Universe by dust or rocks or even spacecraft. While this doesn't solve the problem of how life emerged, it does open an interesting avenue of investigation into how living systems could survive such a journey.

Many microbes show remarkable adaptability, often due to collaboration with other organisms, to environments where more complex life forms would find it difficult to live. The study of such microbes, also termed extremophiles, can provide clues towards answering the broader questions about how life originated in the turbulent conditions prevalent on early Earth, the kinds of off-world environments where life could also have emerged, as well as whether terrestrial life in fact originated somewhere else.

Early Earth was an extreme place compared with conditions today, and yet populations of extremophiles were able to thrive there and continue to inhabit niche regions of our planet to this day. One way to approach imagining life beyond Earth is to study organisms that live in some of the harshest environments on our planet. Extremophiles are organisms able to live or thrive in extreme environments, with temperature, pressure, radiation, salinity or pH beyond what we would consider 'normal'. A range of extremophiles have been able to survive significant shifts in Earth's climate over the past few billion years, and there are lots of them, providing us with a vast array of information as to what life might look like in the challenging environments off-world.

Some extremophiles include microbes found living in highly acidic sulphur springs at nearly boiling temperatures, with some able to be revived after being heated to 420 degrees Celsius, and also at extremely high pressures kilometres below the ocean's surface in the Mariana Trench – temperatures comparable to the surface of Venus and the pressure several times higher.

Following the detection of microbes living in freshwater lakes a kilometre under the ice in Antarctica in 2021, the celestial bodies we believe support water under their frozen exteriors are more enticing than ever in that they may also support life in those waters. Upcoming missions to the moons Europa and Enceladus and also the dwarf planet Ceres may reveal evidence of life there in the coming few years. In contrast to life in the Inner Solar System, which, due to the high volume of material exchange between the first four planets would highly likely be related to us, life beyond Mars may be completely different.

How different, you may ask? Perhaps we don't have to look too far to begin to imagine. Take octopuses, for example. Cephalopods, the most intelligent, mobile and largest of all molluscs, are thought to have emerged on Earth some 530 million years ago. In this class, squids and octopods brought their shells inside their bodies around 276 million years ago. The nervous system of an octopus isn't centralised, like most vertebrates; about one-third of its neurons are found in the brain, with the remaining two-thirds spread throughout the body. This means that octopuses can make fast decisions at the point of contact. The genome of the octopus shows an astounding level of complexity – with 33,000 protein-coding genes (65 percent more than humans) – compared with what we know about other species that have evolved over similar timeframes. Some claim that this rapid development, as compared with observed baseline rates of evolution of terrestrial life, could have been due to a virus. But the claim that a large brain, sophisticated nervous system, independent eyes, flexible bodies and the ability to switch colour and shape for instant camouflage were coincidently formed by a viral infection, and so rapidly, seems tenuous.

What if these otherworldly creatures did in fact arrive from the Outer Solar System? We know that moons with icy crusts and subsurface oceans are candidates for life out there, and Snowball Earth may have been a comfortable home away from home at this time for life from the far reaches of the Solar System. Is it possible that cryopreserved, and fertilised, cephalopod eggs ejected from beneath the surface of an icy moon in the Outer Solar System were delivered into our ocean several hundred million years ago – a time when around 100 times as many meteorites struck the Earth per year as compared with today? Perhaps their most amazing feature, and differing significantly from typical terrestrial life, is that octopuses routinely edit their own RNA to adapt to changes in their environment; this would certainly have been a useful skill if they had relocated to a new world.

Cetaceans are another example of creatures living in our oceans, the intelligence of which we are only beginning to understand. Recent applications of machine learning to vast data sets of recordings of dolphin whistles and whale song may enable us to decode these complex communications in the near future. My dad wrote a book inspired by my participation in the Mars One Project called *Messages from the Deep*. In it, the main characters travel to an exoplanet resembling Earth, and their ability to understand cetacean languages leads to the discovery that the dolphins there have deep connections in space and time, and very important messages to share with us back on Earth. What might we learn from these ancient beings if we could communicate with them?

While the questions of whether fertilised eggs could be viable after a journey through space and what the cetaceans who have been swimming in our planet's oceans for 50 million years may have to say to us remain unanswered, of the many extremophiles found on Earth it's worth focusing on those that we already know can survive the most extreme of environments: space.

A whole range of microbes are known to be able to survive the low-temperature, radiation-vulnerable vacuum of space. These include fungal spores and also tardigrades: eight-legged micro-animals and some of the most resilient creatures known. In a recent season, the writers of *Star Trek* envisaged an organic propulsion system using fungal spores to 'jump' across a mycelial network; an intergalactic ecosystem spanning the entire multiverse. Tardigrades are navigators in this network, through their symbiotic relationship with the spores and their ability to incorporate them into their DNA. While such life-based propulsion systems remain for now in the realm of science fiction, these kinds of extremophiles may share characteristics with the kind of organisms that could exist on other planets or moons in our Solar System, or perhaps beyond.

In fact, tardigrades may already be living – or rather hibernating – beyond Earth. As mentioned earlier, in 2019 Beresheet, a robotic lander developed by a private company, carried human DNA samples, 30 million tiny digitised pages of information about human society and culture and 1,000 illicit tardigrades to the Moon. The mission crash-landed on the lunar surface, and it is unknown whether the archive, or the tardigrades, survived. Dehydrated tardigrade adults can survive for minutes at temperatures below negative 250 degrees Celsius (lunar temperatures can drop this low), and for even longer periods in high-energy radiation at levels hundreds of times greater than a fatal dose for humans.

Even if they did survive the impact, however, the lack of liquid water, oxygen and food means they would be in an extreme dormant state called cryptobiosis. While reproduction would not be possible, tardigrades can prevail like this for decades, coming back to life in just a few hours when re-exposed to water.

While we enjoy the highs, like new knowledge, and lows, including crash landings, of space exploration, all in our mission to understand ourselves, fungi may have beaten us to it. Let's go back a bit. Constituting a fascinating kingdom of their own, fungi are a particularly mysterious form of life. Our estimation of when fungi emerged keeps shifting back in time, the more we learn. Genetic analysis indicates that all fungi are descended from a most recent common ancestor at least 1.3 billion years ago. Billion-year-old fossils found in the Arctic region are further evidence for the early appearance of fungi. Furthermore, 2.4-billion-year-old basalt in South Africa containing filamentous fossils that resemble the root-like networks characteristic of fungi may push back the origin of fungi to more than 2 billion years ago; if so, fungi could even have been the first eukaryotes on Earth.

It is probable that the earliest fungi lived in water, and had flagella to propel them through liquid environments. Fungi were among the first complex organisms to move to land, preparing the way for plants and animals to follow by producing what would become soil from their mining of rock for nutrients. Fungi can also pioneer their way out of extreme environmental shifts; for example, cave fungi may have helped life emerge from Snowball Earth. The theory is that, if sufficiently widespread, these microbes could have accelerated chemical weathering, in particular phosphorous production, thus stimulating marine life like algae, which in turn catalysed atmospheric oxygenation, and the explosion of animals using it as a resource in the aftermath of the global glaciation around 540 million years ago.

By almost 500 million years ago, fungi had formed a symbiotic relationship with liverworts, some of the earliest plants, and since then with a huge range of other organisms: lichens are symbioses between fungi and unicellular algae; mycorrhizae are symbioses involving fungi and the roots of plants; even insects enter into relationships with fungi. Many fungal mutualisms are driven by the ability of the fungus to decompose organic substrates that are inaccessible to its partner, thus playing a fundamental role in nutrient cycling and exchange on a global scale.

Fungi facilitate vast communication networks in living systems like forests, practising inter-species cooperation through resource distribution. My mom has been telling me for years that plants communicate with each other (and with her), and more recent research has shown us one way in which this does indeed take place. Using their branching membranes, fungi build a communication network called the mycelium that connects individual plants, often throughout the entire ecosystem. Mycelium, a root-like network of fungal threads, distributes nutrients including sugar and water, even delivering chemical signals between plants. A single fungus in a forest in northern America has grown to a size of nearly 1,000 hectares; by some estimates almost 9,000 years old, it may be the largest and oldest living organism ever discovered.

Lichens, the symbiotic team-up of algae and fungi, were first identified in the late nineteenth century and cover 8 percent of the surface of the planet; they often live in places where no other life can survive, including space. Above 5,000 metres on Africa's highest mountain, Mount Kilimanjaro, these are the only life forms to be seen, besides the heavily breathing humans far out of their comfort zone with half of the amount of oxygen available at sea level. Fungi have been found to be able to process plastic, heavy metals and even radioactive material, so could be used to reverse pollution in our environment. In the wake of the radiation dose received around Chernobyl in Ukraine, fungi have been seen sprouting from wood inside deserted living areas and also detected inside the nuclear power plant, seemingly feeding on the radiation there by radiosynthesis: using the pigment known as melanin to convert radiation into chemical energy. These fungi could help protect people from radiation, for example as a radiation shield for missions into space. But is that perhaps where they come from?

Fungal spores, the microscopic biological particles that enable fungal reproduction, are able to survive the vacuum of space. Of all microbes tested, only the symbiotic eukaryotic associations of alga and fungi in lichens were fully viable after two weeks in outer space, exposed fully to the radiation, vacuum, temperature extremes and microgravity present there. The time spent in the high levels of ultraviolet light found beyond the ozone layer is a limiting factor – it seems that UV light is what did the lichens in, in the end. However, when populations of microbes, including bacteria, fungi and lichens, are shielded against radiation by being embedded in clay or in artificial meteorites made of meteorite powder, they can survive in space for years. These findings support the possibility of interplanetary transfer of microbes within meteorites.

Fungi can play a role in so many different areas: they can be used to produce building materials in a field called mycotecture; are a source of antibiotics like penicillin; provide a wide range of nutrients; and play a critical role in managing waste. All in all, fungi have been a great companion to life on Earth. Given their resourcefulness and their hardiness, it stands to reason that we ought to factor them in as companions to life *off* Earth. Judging by the diverse fungal species found living both inside and outside various of our space stations in the past, for any future steps into space we will be taking fungi with us, whether as part of the mission design or not.

Perhaps the most striking characteristic of fungi is the psychoactive effect on mammals (including humans) of consuming psilocybin and psilocin-containing mushrooms. So-called magic mushrooms have been shown to boost brain connectivity, as well as to be effective in treating mood and substance use disorders. Psychedelic experiences are described as revealing fundamental truths about the world, like the interconnectedness of our Universe and everything in it. The earliest evidence of their use dates back thousands of years.

Ethnobotanist Terence McKenna proposed the stoned ape theory, claiming that the transition from *Homo erectus* to *Homo sapiens* and the associated cognitive revolution was triggered by the addition of psilocybin mushrooms to the human diet a few hundred thousand years ago. Given that tracking animals was a central activity for early humans, and that mushrooms bloom in animal dung, this could indeed have been a contributing factor to human evolution. An even more fringe theory is that fungi did this on purpose.

Given their level of adaptation to the harsh environment of space, we may wonder whether fungi have a history there. While fungi can produce complex structures, rockets may not be one of them. As animal behaviourist Kurdt Greenwood has pointed out to me, perhaps they have driven the evolution of complexity of life on Earth, and finally humans, precisely in order to go on an interplanetary trip. Perhaps not their first.

Life on Mars?

Given that we remain for now in the direct vicinity of our home planet, the remote search for life beyond Earth entails, if not the observation of life itself, the detection of evidence that living processes are taking place. Life requires resources to survive, and by-products of the utilisation of resources can provide clues as to where to look for living organisms. While layers of ice, dense atmospheres and years of travel mean that much remains hidden from our view, there is a place beyond Earth visible to the naked eye with which, through remote exploration, we are becoming increasingly familiar. In 2020 we could even say there was traffic in this direction, if three missions launched in the same month constitute traffic. On arrival, we also flew the first off-world drone to have a better look around. If the truth about where we come from is out there, it may be closer than we think. Besides being our next-door-neighbour planet, what is it about Mars that makes us think there's something there, that has captured our imaginations, inspiring books, films and music in a way that no other body in the Universe has?

The word 'planet' is from the Greek word meaning wanderer. With the far more distant stars in the night sky appearing fixed in their relative position to each other, early humans would have observed that the much closer planets of the Solar System appear to wander about. Mars was first observed with the newly invented telescope by astronomer Galileo Galilei in 1610, and within the next century astronomers had determined the planet's rotational period, just thirty-nine minutes longer than Earth's twenty-four-hour cycle, and its axial tilt, meaning that Mars has seasons. Much interest in Mars was sparked by astronomer Giovanni Schiaparelli's claim in 1877 that he had observed oceans, continents and channels on the surface. The Italian word *canali* (meaning channels) was misinterpreted as implying artificially built canals, and speculation about engineering works and irrigation systems constructed by a civilisation of intelligent aliens indigenous to Mars abounded.

In the early 1900s, public imagination was further fuelled by films depicting intelligent life on Mars. In 1910, *A Trip to Mars* was produced by one of inventor Thomas Edison's film companies, and up until the 1960s more films featured Mars and advanced extraterrestrials living there. With bated breath, the Mariner 4 flyby mission was launched to Mars in 1964. Disappointingly, the images it sent back revealed the surface of Mars to be a dry, barren place, devoid of oceans and obvious signs of life.

Schiaparelli was right about the channels, however; Mars was once much warmer and wetter than it is today. Geological structures like valleys, deltas and lake beds are clear evidence that great bodies of water once existed there, billions of years ago, prompting the question: did life once exist in these bodies of water? We needed to get closer to answer that question.

Our first on-the-ground knowledge of Mars was provided in 1976 by the Viking missions. Viking 1 and Viking 2 were twin space probes that made history by being the first to land safely on Mars, revealing evidence that there had been water flowing on a younger Mars, contrary to the present arid conditions, and also measuring the abundances of the atmospheric constituents. The Viking orbiters sent back images of a place called Cydonia, in the Arabia Terra region of Mars, the area where the southern highlands meet the northern plains. This is characterised by wide, debris-filled valleys and isolated mounds of various shapes and sizes seemingly formed by erosion. The region elicited great excitement when the Viking spacecraft returned a low-quality image of a large rock formation resembling a human face as well as an array of structures that some claim are artificial. The controversy around whether the structures are natural or not is claimed to be solved by higher-resolution images obtained from orbit a few years ago.

The human tendency to see familiar patterns like faces in inanimate structures and random patterns, or pareidolia, is cited as the reason for the confusion, but personally I'd appreciate the opportunity to go and have a look myself. Unfortunately, we're not quite there yet.

No less controversial than the alleged artificial structures on Mars was a positive outcome for the detection of extant microbial life on its surface; experiments on both Viking landers, over 6,000 kilometres apart, yielded 'similar, repeatable, positive responses'. An explanation could be that a yet unknown non-biological component like an oxidant in the Martian soil triggered the response. In any case, it is widely agreed that the stringent, but now outdated, decontamination procedures applied to the Viking missions were likely insufficient. The detection in the sanitised environments where space missions are prepared of microbes with enhanced DNA repair capabilities, enabling increased radiation resilience, may in fact indicate that such facilities can act as natural filters serving to enhance the likelihood of microbes enduring a voyage to Mars. The potential for such contamination of distant celestial environments with terrestrial life prevents us from knowing for sure whether or not the Viking missions detected local life there or perhaps terrestrial hitchhikers; yes, microbes including fungi may already have beaten us to Mars.

While the detection of life there remains controversial, what we can agree on is that the Cydonia region was once a place where large volumes of water flowed; some suppose it was a coastal region at the time, around 4 billion years ago, when Mars had oceans. The widespread presence of hydrogen detected from orbit in the region indicates a previous but also potentially present abundance of water below the surface. More recently, dark streaks along slopes in the Valles Marineris region near Mars' equator were seen to grow more prominent during the warmest times of the Martian summer and then gradually fade as an annual occurrence. Since pure liquid water would boil in the pressure and temperature conditions on the surface of Mars, it has been proposed that these are the residues as brine (water with dissolved salts) flows downhill in warm conditions and then evaporates. The source of the water, however, remains unknown.

Given that water is vital for all life here on Earth, and also taking into account the large amounts of debris exchanged between the two planets, particularly at the time when they both had vast oceans, it is not inconceivable that life on Earth was first seeded by a Martian meteorite containing space-resilient microbes of some sort that landed here. Several thousand tonnes of space debris fall to Earth each year, with around a percent of all identified meteorites found to be Martian. And indeed, an intriguing piece of ancient Mars has already been found.

Antarctica, with much of its surface covered with ice, lends itself to the search for rocks that have arrived from elsewhere in the Solar System; the so-called Allan Hills 84001 meteorite fragment was found in Antarctica in 1984. The two-kilogram rock contains gas bubbles trapped inside, and the composition and isotope ratios of the gas provide convincing correlations with the atmosphere on Mars and thus evidence for where the meteorite originally came from. The fragment is believed to be the oldest known piece of Mars here on Earth (out of a few hundred meteorites confirmed to be Martian), estimated to have crystallised from molten rock around 4 billion years ago when Mars had liquid water on its surface. It is thought to have been blasted out from the Valles Marineris region around 17 million years ago by a meteorite impact there, and to have fallen to its resting place in the south of Earth around 13,000 years ago. While this particular rock was a few billion years too late to have seeded life here, it is evidence of the possibility of such a life-transfer event. And potentially evidence of life on Mars. Let's see why.

After more than a decade of analysis, in 1996 scientists announced that they had found several features suggesting the presence of microfossils in the meteorite, including ones of bacteria-like objects, suggesting that these organisms originated on Mars. The potential detection of life on Mars elicited the announcement by US president Bill Clinton: 'Today, rock 84001 speaks to us across all those billions of years and millions of miles. It speaks of the possibility of life.'

However, some of the scientific community ultimately rejected the hypothesis, claiming that the unusual features in the meteorite had been explained without requiring the presence of life. In spite of the remaining controversy over whether or not the rock contains evidence of life, the scientific and public attention generated by Allan Hills 84001 is considered pivotal for the development of the science of astrobiology – the study of the origins, evolution, distribution and future of life in the Universe – which, among other things, needed to come up with a list of requirements for the identification of extraterrestrial life.

In 2012, the Curiosity rover successfully landed on the surface of Mars and set about its mission to determine whether the Red Planet could ever have supported life. The robot is about the size of a car, and kitted out with scientific instruments used to study the planet's climate and geology, as well as focus on its main objective: the search for evidence of life.

Curiosity landed in an area called the Gale Crater, formed when a large space rock hit the surface of the planet. Later on, this crater was filled with liquid water, and layers of sediment built up over time into the sides of the mountain. Within this sediment we can read a chronology of the early history of Mars and determine which periods had conditions that could support known life: we have learned that Mars was probably habitable for at least tens of millions of years, around 3 to 4 billion years ago, when Mars was a much more Earth-like place than it is today. But has Curiosity also found evidence that life has prevailed even in the present-day conditions on Mars?

Carbon dioxide transitions between solid and gas states under the kinds of seasonal variations in temperature and pressure taking place on Mars; so much so that overall atmospheric pressure can fluctuate by as much as 30 percent owing to the amount of carbon dioxide in the air. When carbon dioxide gas freezes over the poles in the winter, requiring temperatures below negative 120 degrees Celsius at typical Martian pressures, the air pressure across the planet drops; this carbon dioxide then evaporates in the spring and summer, resulting in an increase in the global air pressure. It has been found that other components of the Martian atmosphere, namely nitrogen and argon, also follow predictable seasonal patterns in the Gale Crater throughout the year relative to these carbon dioxide fluctuations. Oxygen, present in trace amounts in the air, however, has been observed by Curiosity's portable chemical lab to behave anomalously – rising in abundance throughout spring and summer, also by as much as 30 percent, and then dropping back to expected levels by autumn. This pattern seems to repeat annually, although varying in the amount of oxygen added each season; however, unlike the carbon dioxide fluctuations, no known chemical process can explain this cycling.

Another mysterious feature of Martian air is the presence of methane. While a number of missions since the first detection in 2004 have reported trace amounts of methane in the Martian atmosphere, without an obvious source for this methane and the periodic spikes in its concentrations we are still none the wiser. Inside Gale Crater, barely observable amounts of methane have also been measured by Curiosity to rise and fall seasonally, increasing in abundance by as much as 60 percent in summer months, and sometimes correlated with the anomalous oxygen fluctuations. Whenever methane has been detected in these observed spikes by Curiosity in Gale Crater, just like the oxygen, it is subsequently removed from the atmosphere by an efficient yet unknown process. The presence in the Martian atmosphere of oxygen, which is almost entirely produced terrestrially by photosynthetic activity, and methane, which is emitted in large amounts by animals and microbially driven organic decay here on Earth, may indicate the presence of life, some unknown chemical process, or alternatively some kind of geochemical process such as volcanism or hydrothermal activity, or perhaps both.

Curiosity is also equipped with a drill, and in early 2022 it was announced that enriched amounts of carbon-12 (life's most crucial isotope) had been discovered in powdered rock that Curiosity drilled from a hole in a rocky ridge in Gale Crater. While on Earth carbon-12 is typically a sign of life, and while life may be the simplest explanation for the fluctuations in oxygen and methane detected in the crater, our as yet limited capabilities to perform experiments remotely mean that, even considered together, this evidence is not sufficient to unequivocally announce the presence of life on Mars.

Building on the capabilities and achievements of Curiosity, the Perseverance rover was launched in 2020 to the Jezero Crater on Mars; a site 3,700 kilometres from Gale Crater that also appears to have once hosted a deep, long-lived lake. At one edge, what was a river delta 3.5 billion years ago containing rock and sediment deposits is thought to be a good location to search for potential signs of ancient microbial life. Perseverance, inspired by and moderately upgraded from its predecessor Curiosity – although, surprisingly, not equipped to analyse atmospheric methane or its isotopes – carries seven instruments to analyse the environment, nineteen cameras and two microphones. Perseverance also carried a small but impressive passenger: the Ingenuity helicopter. Flight in Mars' thin atmosphere is challenging; Ingenuity weighs just 1.8 kilograms on Earth. Originally intended to make only a handful of flights, Ingenuity completed seventy-two, covering 17 kilometres, before being permanently grounded in 2024 due to rotor blade damage.

Tragically, Namibian spaceflight engineer Japie van Zyl was not alive to see Ingenuity, his and his team's brainchild, flying and communicating autonomously in what was the first powered flight on another planet. He passed away in August 2020 soon after Perseverance launched the previous month. As the director of Solar System Exploration at NASA's Jet Propulsion Laboratory since 2016, he was responsible for multiple missions to the Outer Solar System. An inspiration to me – I'm still working on some of the ideas we discussed when I met him a few years back – and to young people in Southern Africa and around the planet, his memory lives on, on both Earth and Mars, and indeed throughout the Solar System.

Besides successfully flying the first drone on Mars dozens of times, Perseverance's goals include identifying ancient Martian environments that could have supported life, searching for evidence of microbial life that may have existed in those environments, collecting soil and rock samples to store on the Martian surface for future sample return, and testing technology to produce oxygen from the carbon dioxide in the Martian atmosphere to prepare for future crewed missions.

While we've spent a lot of time and energy searching for life on Mars, and while the evidence is compelling, the understandably cautious way in which our robots do experiments 200 million kilometres from Earth has not resulted in any conclusive results, or at least any public announcements of such results. The most interesting outcome of future investigations on Mars, from the perspective of understanding the nature and origins of life, would be to find life forms completely distinct from those on Earth: a second unique data point. However, a more likely feature of life there, given the meteoritic exchange between our planets, is that it would be related to life on Earth; DNA-based, but of a different evolutionary branch since a split potentially billions of years ago. All our experiments are necessarily based on looking for life as we know it, and our intrigue at the fluctuation of things like methane and oxygen is similarly based on our knowledge of terrestrial microbial activity.

If we are to increase our level of understanding of where we come from, as well as understand on a deeper level what life is, then studying the network of living systems on Earth is only going to get us so far. We need at least one more data point, one more example of biology elsewhere, so that we can begin to think about the fundamental principles underlying the phenomenon of life, as well as the kind of conditions where it can exist.

While we continue the remote search for evidence of life beyond Earth, are there other avenues of investigation?

The search for extraterrestrial intelligence

The signatures of microbial life are difficult to distinguish conclusively from complex non-biological processes, especially from millions and millions of kilometres away. However, perhaps advanced extraterrestrial life may be active on scales large enough for us to detect from Earth, and perhaps even closer to home than we imagine.

In a 2020 report, the Pentagon simultaneously acknowledged the existence of and denied US involvement in a range of unidentified aerial phenomena (rebranded from 'UFOs') reported primarily in the US since the 1940s. With no other parties stepping in to take responsibility, many have leaped to the conclusion that these are the result of intelligent alien visitation. And yes, if there were events that would prompt an interstellar intervention, they may well have been 'Little Boy' and 'Fat Man', devastatingly deployed on hundreds of thousands of citizens living in Hiroshima and Nagasaki in 1945: a demonstration of our new-found capabilities in manipulating the protons and neutrons inside the nuclei of atoms, where more than enough energy to cause a planetary nuclear apocalypse lies hidden. 'I am become death, the destroyer of worlds,' said the head scientist, J. Robert Oppenheimer, quoting the Hindu scripture the Bhagavad Gita to describe his feelings when 'it worked'.

Are we alone?

If a distant civilisation were intelligent enough to travel far enough to visit us, it may find it wise, and surely not difficult, to remain undetected by us.

Of course, instead of waiting for a visit, we could – and do – actively look for intelligent life beyond Earth. Encoding and sending information using light is the fastest way that we know of to communicate, so we may guess that if civilisations do exist elsewhere, they might do the same. Longer wavelengths travel better through Earth's atmosphere even in cloudy conditions, and also through things like interstellar dust, so the radio part of the light spectrum, with wavelengths between a millimetre and thousands of kilometres, is typically where organisations like the SETI (Search for Extraterrestrial Intelligence) Institute, Breakthrough Listen (the search for intelligent extraterrestrial communications) and others are looking. To receive these longer wavelengths, some almost the diameter of Earth, we need a large collecting area.

The Square Kilometre Array (SKA) radio telescope is currently under construction in South Africa, African partner states and Australia. The initially proposed receiving area of more than a square kilometre has since been expanded, and will enable the observation of a wide range of radio frequencies with better sensitivity than any other radio instrument we have ever built, and a deeper look with more resolution than ever before into our Universe. Among the range of more traditional astronomical applications for the instrument is the search for extraterrestrial communications.

Out of This World and Into the Next

In 2018, in Bremen, we celebrated the announcement of South Africa's radio astronomy collaboration with the Breakthrough Listen Project: to investigate 1 million stars in the search for intelligent life beyond Earth. And then, in 2019, Breakthrough Listen detected an interesting signal coming from the direction of Proxima Centauri. This was immediately exciting: the curious repetitive signal at 982 megahertz might be a signature of alien technology, since no known natural process can account for the repetitive, single-frequency signal. As of yet, no human technology has been identified as the source of this potential technosignature, but reports say that these signals are likely interference that we cannot fully explain, and that further analysis is underway. As of the end of 2020, follow-up observations had failed to detect it again, but the signal, named BLC1, does have its own Wikipedia page. 'No, we still don't understand what this was,' said Pete Worden, the director of Breakthrough Listen, when we bumped into each other in California in 2022.

Another place we could look for intelligent extraterrestrial communications is a part of the Universe that has until now remained hidden from view: the sub-10-megahertz portion of the electromagnetic spectrum. Earth's outer atmosphere absorbs and reflects signals with wavelengths greater than around 30 metres, which corresponds to a frequency of 10 megahertz, and thus our observation of the radio sky at longer wavelengths is best performed from beyond Earth. Due to physical shielding from the Earth and periodically the Sun, low-frequency lunar farside-based radio astronomy could reveal new information about the early Universe prior to the formation of the first matter 380,000 years after the Big Bang. We could also discover unknown unknowns. If nearby intelligent life chose a frequency band in which to communicate, perhaps one not detectable from Earth would have had its advantages; the search for intelligent extraterrestrial communications below 10 megahertz is something we have not achieved yet (more on the lunar-based Africa2Moon radio telescope coming soon).

There are, however, a few problems with looking for electromagnetic signatures of intelligence: the Universe is really, really big and light travels fast, but not infinitely fast, meaning that even light takes millions of years to travel between galaxies. Then there's also the possibility that civilisations only use this method of communication for relatively brief durations. If the 4-billion-year history of life on Earth is a day, then the 300,000 years during which humans have been around is equivalent to less than a second. And the time since inventor Guglielmo Marconi sent the first radio signal across the Atlantic Ocean in 1901 is far less than a blink of the eye. How much longer we continue to send such signals is anyone's guess, whether due to self-annihilation or developing an entirely new communication technique; regardless, it probably won't be that long in the larger scheme of things. Putting all this together, it's not likely that anyone would be in the right place, at the right time to receive radio communication signals from someone else at another time and place elsewhere in the Universe.

Searching for signals from intelligent extraterrestrial life is not the only way in which we might detect alien life. After all, whether or not a civilisation is using radio communication, perhaps in stealth mode, if they do exist in the physical world they will likely require resources and be involved in some kind of industry; the more advanced, the more activity. I attended a striking presentation by astrophysicist Beatriz Villarroel where she described her recent work: the search for changes in existing astronomical survey data, the line of thinking being that objects in the night sky that suddenly appear or disappear without known explanation could be evidence of technologically advanced civilisations that are utilising resources or building structures on astronomical scales.

Given that we've yet to become advanced enough to build structures on astronomical scales, we are left to guess at what kind of structures these might be. One particular idea which has since gained notoriety, particularly in the science fiction-writing world, was inspired by the 1937 science fiction novel *Star Maker* by philosopher Olaf Stapledon, and formalised and popularised by mathematical physicist Freeman Dyson in his 1960 article 'Search for Artificial Stellar Sources of Infrared Radiation'. The structure proposed, a so-called Dyson sphere, went something like this. Imagine an advanced civilisation whose energy requirements exceed its home planet's available resources. The so-called Dyson sphere is a hypothetical megastructure that this civilisation builds around a star to capture the majority of its power output. The infrared heat that unavoidably escapes from any system of energy collectors would be a signature of a Dyson sphere. While so far we have not detected any evidence of extraterrestrial megastructures, the constantly improving resolution and sensitivity with which we are observing the Universe around us may yet reveal something.

Analysing the light arriving to our atmosphere is currently one of the best ways in which we can gather information on what is happening beyond our planet. However, when we make observations from Earth, we are always looking into history because of the finite speed of light. What we are in fact seeing is what existed at the time when the carrier of the information, the light, left its previous distant location. The Cosmic Microwave Background gives us a picture of what the early Universe was like billions of years ago; light from the nearest stars gives us data on the state of affairs there several years ago. So even if we intercepted any alien communication, and even if we could decipher it, it would be old news. Likely, very old news.

A civilisation capable of harnessing the entire radiation output of its star, perhaps using a Dyson sphere, would be Type II on the Kardashev scale, proposed by astronomer Nikolai Kardashev in 1964 for measuring a civilisation's level of technological advancement according to the amount of energy it uses. We are still developing towards a Type I classification, since despite questionable proposals to harvest solar energy in space and beam it to Earth using microwaves, we currently use thousands of times less energy than the total solar radiation impacting our surface: in one hour, enough sunlight to power the electricity needs of every human being on Earth for a year arrives in Earth's atmosphere. Even photosynthesis does better than us on the Kardashev scale, using up to a couple of percent of total incoming sunlight globally. Where does all of this leave us?

More things than dreamt of in our philosophy

We've been on a journey from the edge of the observable Universe at the beginning of time to the first energy, matter and resultant structures in the Universe, all the way to our own Galaxy and Solar System, and finally to the story of Earth and the life that emerged here – all in a quest to find out where we come from. A definitive answer to this question would be a scientific theory of life. This would require a rigorous description of how it emerges from the inanimate matter of which it is made, as well as a bunch of experiments showing how, according to this theory, life can be created in the laboratory, in conditions resembling those of early Earth, and potentially in other kinds of simulated environments also.

Finding evidence of life beyond Earth would be the most profound discovery in our history, advancing our perspective of reality and ourselves in unimaginable ways. If there is life beyond Earth, there are two options: either we are related or not. Given the distance from us and the vastly different conditions, if there is life swimming in oceans beneath the ice in the Outer System it is possible that it's completely distinct from ours. With our exploration of space having been, so far, limited to our Solar System, finding evidence of a second separate lineage of life orbiting our Sun – while not immediately telling us where *we* come from – would be the holy grail of data points in understanding life itself: a second piece of the puzzle of a general theory of life, as well as a strong indication that life is far more abundant in the Universe than we may have imagined. An alternative giant leap in understanding our origins would be the detection of organisms genetically related to terrestrial life. And with both of our neighbours thought to have once had oceans, we may not have to search too far. We believe Mars had vast bodies of surface water even before Earth became hospitable to the earliest terrestrial life: our extraterrestrial ancestors may lie just a few light minutes away, preserved, or perhaps still living, beneath the ancient rusty crust of our next-door neighbour. In the absence of such evidence, for now, we are left to theorise.

Understanding how one of life's oldest processes in the simplest of organisms operates on a quantum level is a far cry from understanding life itself, but it is a start.

Quantum biology's greatest achievement would be a contribution to a general theory of life. In spite of impressive work in the field, the modelling from a physics perspective of much of the kind of complexity we see in biology remains beyond reach. Perhaps a quantum computer could help (we'll look at quantum computing in more detail later), but we're still years away from a system capable of simulating even simple life. And many gaps in our knowledge, beginning with the formation of the simplest molecules constituting life, are yet to be filled.

Physicist Richard Feynman observed: 'what I cannot create, I do not understand'. While our efforts at a theoretical understanding of life are ongoing, perhaps we will get lucky in the lab; perhaps experimental work will shed light on the issue of how life emerged in the first place. Although the intriguing Miller–Urey experiment in the 1950s showed the spontaneous production of a whole range of amino acids in the laboratory, we're still far from striking life into the dead with electricity in the manner that Mary Shelley envisaged in her famous work *Frankenstein* in 1818.

I attended an astrobiology conference in Chicago a few years ago. There I met multidisciplinary scientist Bruce Damer, who, after telling me that he has psychologist Timothy Leary's ashes in his computer museum, explained that ideas and answers to questions can be revealed, conjured up, through concentrated meditation; that he has seen how life formed. Bruce, and a growing community of origins of life researchers, envision a compelling account for an origin of life on land in a twenty-first-century version of Darwin's 'warm little pond'; an account increasingly supported by evidence and experiment.

The model begins in shallow pools of water on the volcanic landscapes characteristic of early Earth, places where source organics from hot springs, meteorites or dust can accumulate and concentrate, as opposed to in larger bodies of water like the ocean where they remain dilute. As demonstrated through laboratory work and at field analogues, wet–dry cycling can result in the formation of biomolecules like RNA, DNA and amino acid chains and encapsulate them into compartments of fatty compounds called lipids, while at the drying edges of pools and with exposure to sunlight, glistening bathtub rings composed of layered membranous structures can form organic building blocks including nucleic acid precursors. Collectively, these complexes then form spherical collections of lipids, or protocells, which cycle through wet, moist and dry phases and undergo the first forms of selection and evolution towards the living cellular world. In support of the hot spring hypothesis, many of the steps in the emergence of the key chemical circuitry and protocells required for life itself have been demonstrated.

In spite of such success, and the beautiful imagery conjured, no experiment has yet been able to synthesise from basic components an object that has the characteristics of a living cell; we have not been able to derive biology from chemistry in the laboratory, and therefore cannot claim to understand life from a scientific perspective.

Trying to understand how life emerged on Earth, how collections of atoms and molecules became organisms in a possibly singular event, is not something current science has been able to explain. In fact, our science seems insufficient to reveal a theory of life; and, with our traditional approach, perhaps it never will. But that is not to say we haven't made some important discoveries along the way. We do know that we contain atoms as old as our Universe, molecules older than our Sun, and share genetics and a common ancestor with all living systems on Earth from a time soon after our planet was formed. Our deepest enquiries into the origins of matter and life tell us that we are in the Universe and the Universe is in us, all the way down to the most fundamental constituents of reality.

But in terms of trying to understand what life is and where it comes from, we remain limited to a single data point in our study of the family of terrestrial life. Our scientific approach is limited; but perhaps by something more fundamental than just the availability of data. As Shakespeare's Hamlet said, 'There are more things in heaven and earth, Horatio, than are dreamt of in your philosophy.'

In my third year of university, for the first time, I made an appointment with my physics lecturer to ask a question.

Are we alone?

I had become interested in quantum mechanics, having learned about the Uncertainty Principle, and how quantum physics implies that the act of measurement always disturbs the object being measured. My question was whether observing something without interacting with it, and therefore disturbing it in some way, was in principle impossible, or whether the issue was due to our own technical limitations. He answered by challenging me to devise a completely non-intrusive observational method.

It didn't take me long to realise that all known methods of observation include interaction, from human exchanges all the way down to the quantum level; a single particle of light – a photon – that bounces off of an atom, telling us about the existence of that atom, involves an interaction between the atom and the light particle, leaving its mark on both. While a single photon may be imperceptible to a human among all the other noise of reality, at a quantum level this constitutes a significant disturbance.

We concluded that a system that never interacts with any other system remains but a philosophical construct rather than an element of reality. The implication that the observer is then inextricably part of the system under observation has fascinated me ever since. I was intrigued by the realisation that the way in which a question is posed can influence the answer, from the level of interpersonal engagement all the way down to measurements performed on single particles. The unique contribution that a creative human mind can have in shaping the direction of scientific discovery is what drove me to continue studying theoretical physics.

I have had the privilege to spend many years in thought, studying the efficiency and robustness of photosynthesis, investigating how things like amino acids form in space as well as in the laboratory, all the while trying to understand life and whence it comes, from the quantum physical perspective of the electrons, atoms and molecules that make up all living things.

I'm not alone in thinking that our current scientific view is incomplete. Despite a century of trying, we cannot reconcile our theory of the very big, general relativity, with our theory of the very small, quantum physics. And in no realm are our shortcomings more obvious than in the biological world, where the complexity baffles not only our minds but the best computers we can build, or even imagine! We are going to need more than current science, and likely more than a quantum computer, to get to the bottom of what life is.

What are we missing? A good start would be to acknowledge that the distinction between 'us' and 'nature' that has been the core doctrine of science, and indeed our modern culture, is already revealed to be flawed at a fundamental level. We are inextricably interconnected with the Cosmos in which we find ourselves and everything in it; our bodies are made of the same energy and matter that has filled the Universe since the beginning of time. But what about our minds? Inventor Nikola Tesla said: 'My brain is only a receiver in the

Universe, there is a core from which we obtain knowledge, strength and inspiration. I haven't penetrated the secrets of this core, but I know that it exists.'

For now, if you can't think of any specific proposals for a scientific revolution or a quantum computer, or how to access this 'core' offhand, don't despair. As I did sitting in front my computer as a postdoc pondering H_2 formation and deciding I needed a new approach. While this is where this side of the story ends for now, to any good story there is always more than one side. Although we are on the threshold of a new era for humanity, a transformation in our understanding of ourselves and our place in the Universe, for now we are still confined to a very limited perspective of the reality in which we find ourselves. The Earth is the cradle of humanity, but, as one of the founders of rocketry, Konstantin Tsiolkovsky, put it, we cannot stay in the cradle forever.

In fact, the countdown is already on: in the next few hundred million years the Sun will approach red giant phase, becoming hotter and brighter and expanding beyond the orbit of Venus in size. In a billion years the Sun will be 10 percent more luminous; our oceans will evaporate and surface temperatures will skyrocket to hundreds of degrees Celsius, conditions under which all life as we know it will cease to exist. While this may sound far off, on cosmic scales this means that the 4-billion-year run of life here is 80 percent done; yes, we are approaching the final stages of our track record on Earth. In the remaining 20 percent of the lifespan of our planet's biosphere we will need to develop the capability to expand beyond home to ensure the continuity of terrestrial life. And some of us dreamers remain dedicated to preparing for a future among the stars. After all, as we gaze up at the night sky and into the reality in which we find ourselves, let's remember that while we are in the Universe, our home, we are not separate from it: the Universe is also in us, right down to the original light and matter produced at the beginning of time. Our ability to explore has advanced such that the establishment and potential detection of life beyond Earth is within reach; we just need to get there and look.

But, you may argue, can't we use technology for that? Sufficiently sophisticated robotics and remote-sensing techniques could, eventually, become more capable than humans for such tasks in hostile off-world environments.

To which I answer: perhaps, but where's the fun in that?

PART II

WHO ARE WE?

Off-world exploration, much of it driven by the desire to know whether we are alone in the Universe, may well lead to the confirmation of life beyond Earth in the coming years: a discovery that promises to advance our perspective of ourselves and our world in unprecedented and unimaginable ways. Another reason to explore is survival: the ability to migrate to and innovate in new environments has enabled the continued existence of our species thus far. Importantly, exploration is also a way to better understand ourselves.

Our journeys into the unknown have defined us a species for as long as we have walked the Earth; driven by natural disasters, climate change and the search for resources, but also by innate curiosity, courage in the face of the unknown and the aspiration to establish new paradigms. These journeys were tough, but we brought with us tools and ideas that led to the creation of societies with new systems of thinking and organisation. It is our ability to travel and adapt that enabled our population to reach a size of a billion people by 1804. And in spite of the First World War and the subsequent influenza pandemic, this number had doubled by 1927, and doubled twice again to reach our current population of 8 billion people today.

As we run out of new spaces to migrate to here on Earth, the global disasters we face are increasingly self-induced; our population size and the extent of our activities across our finite planet have destabilised our natural life-support system, and in the midst of the resulting mass-extinction of our fellow terrestrial species, our society faces challenges like epidemics, extreme weather and conflict over resources. An important lesson we can learn from our hunter-gatherer ancestors is that a community able to migrate in times of great upheaval has a greater chance of survival than one with a fixed notion of home. While there may be no (current) urgency to evacuate the planet, harsh and resource-constrained off-world environments are great places to develop solutions for resilience in the increasingly unpredictable conditions here on Earth, as well as to prepare for more distant migration in our future.

Expanding beyond Earth into the Solar System is not the final frontier, but rather just the beginning of our journey into the stars; yet many people find venturing off-world unfathomable. In fact, there are striking parallels between this voyage and many others throughout our entire history. If we can imagine a time before Google Earth and Tripadvisor, and what it was like to set out towards the horizon to seek new lands, then we can begin to understand that travelling beyond Earth is no different. Off-world exploration missions will take us further from home and with more sophisticated cargo than ever before. But we actually have far more detailed knowledge of what we can expect to find there than the majority of our ancestors did as they explored our home planet, transforming unknown lands and oceans from barriers into pathways. This is what we have always done as humans and what we will continue to do: we observe, we dream and we expand our horizons through the realisation of these dreams.

4

EXPLORERS

The day cosmonaut Yuri Gagarin bravely hurtled around Earth in the first ever craft to take a human into Earth orbit was arguably the beginning of human space exploration. Going back even further, however, palaeo-anthropologist Francis Thackeray, former director of the Institute for Human Evolution in Johannesburg, makes the case that the first steps towards space exploration were taken right here in my home country, South Africa. Here there are multiple cave sites that have been excavated by archaeologists for about 100 years, generating a range of insights into human prehistory. For example, the controlled use of fire was inferred from burnt bones discovered at Swartkrans in the Cradle of Humankind World Heritage Site. The bones had been heated to temperatures higher than that of naturally occurring grass fires. Other evidence was obtained at Wonderwerk cave, 30 metres in from the entrance. The fires were probably used in the same place, over and over again. The people responsible were possibly of the species *Homo erectus*, the first human ancestor to spread from Africa throughout Eurasia, existing between around 2 million years ago up until about 300,000 years before present. Evidence of the controlled use of fire at least 1 million years ago, in Africa, is the earliest technological precursor to rocket propulsion. Wonderwerk means miracle in Afrikaans.

Our origins: forged in fire

During the COVID-19 pandemic lockdown of 2020, my partner at the time, Kurdt Greenwood, and I built a cabin deep down in a valley in the ancient indigenous Tsitsikamma forest. We lived there over the first half-year of lockdown, the only access being a 1.2-kilometre rugged footpath down a 300-metre descent and across a river, the route the two of us carried down a couple of tonnes of supplies to build the cabin. It was a time to contemplate resources, the most important of these being shelter, power, water and food; if we did not collect kindling from the upper slopes of the valley during sunshine, during rain we would have no hot water or cooked food. Collecting water and driftwood from the river, and medicinal plants in what appeared to be ancient terraced north-facing gardens above our cabin, I had the feeling of being part of a long lineage of humans living beside this ancient stream.

We spent a lot of time exploring the area, and one day Kurdt discovered a collection of seashells in the compacted ground under a rock overhang beside the river. This is around 13 kilometres, or a six-hour trek, upstream from the coast where, at the river mouth, one of the largest shell midden deposits on the planet is located in a rock shelter; home to the earliest people of Southern Africa, the San, for over 12,000 years, and a place I've been visiting with my dad to look for stone tools since I was a child. A collection of seashells in compacted ground so far from the coast could mean just one thing.

We returned to the site about a kilometre upstream from our cabin the next day with brushes and buckets and started digging. Kurdt's hole revealed pebbles carved into fishing sinkers and long bone hooks, stone cutting blades and smooth grinding rocks. As a survival expert and animal behaviourist he was hopping with excitement, theorising that the tool maker fished for eel in the deep pockets of the river, and immediately planned to try the technique himself.

The hole I dug right next to his contained nothing at all, so I stepped back to look around. I was drawn to the very back of the overhang, where some fine, dry sand lay in the shadow. I crouched under the overhang and brushed away slowly at the surface. I found a piece of an old bottle, then a pottery shard, then beneath that a collection of crystals and small animal bones, and then suddenly a round, hard shape appeared. A sense of gravity fell upon me. Even as we discussed whether it was a stone, a bulb, a baboon skull ... I knew it was someone's story we were uncovering, a voice that had last spoken thousands of years ago, in another time. We alerted local archaeologists, who confirmed from the tools with the skeleton, as well as the position in which the body was laid to rest facing east, that it is a San burial site. We await the dating of the site, which could be anything between a couple of hundred to tens of thousands of years old.

The night of the discovery, the forest was so close as I closed my eyes to sleep; the deep, damp stillness of the trees, the gurgle of the stream over the rocks, which often sounded like women and children laughing upstream, as they had done in this valley for at least 10,000 years. With some trepidation, I asked that the person whose remains we had uncovered, if unhappy about us doing so, would come to me and let me know. I fell uneasily asleep into a strangely clear dream: I saw a time-lapse of layers of atoms and molecules accumulating through life and weather on the surface of the Earth; I saw that the person who had once lived in this body was no longer attached to the layers of matter left behind. I was told that I was free to sift through this record of what was, if I so desired. I felt more at ease for the rest of our excavation. While so much has changed since this person last roamed the Earth, holding the tools that they used to navigate this same environment so long ago, I felt the unity of all humanity through the ages; a continuity in the story of who we are.

While there are a number of contenders for the region where the first *Homo sapiens* emerged in Africa, genomic analysis of people alive today indicates that the San people of Southern Africa are the most genetically diverse of any living humans studied, suggesting that ancestors of the San were the first modern humans to emerge on Earth. These early humans, who gave rise to all 8 billion of us alive today, proceeded to migrate across the habitable surface of our entire planet, surviving natural global cataclysms on scales we have not seen in modern times that shaped their communities and way of life. That's us! But what is it exactly that makes us uniquely human?

In a course I took on human evolution, a welcome change from the hours of calculation that physics involved, we wore lab coats and measured things like femur lengths and occipital ridges to identify the genus and species of various hominin fossil replicas laid on the desk. The fossil record shows that modern humans are characterised, among other things, by our hip and leg adaptations to walking upright, and our skulls that house our large, complex brains. These physical traits, together with our innate curiosity and desire to explore, gave rise to our development of tools, culture and language; humans are creative. We are also highly social and tend to live in complex communities; humans are collaborative. Our curious, creative and collaborative nature has advanced our ability to explore and adapt. But how did these characteristics come about?

Life requires resources, and living things access these resources from the environment around them. Sudden changes in external conditions pose a challenge to organisms, which then either adapt or die off. Environmental instability has always been a key driver of the evolution of life on Earth, often causing numerous species to go extinct, but also creating the conditions that make evolutionary innovations necessary, thus playing a fundamental role in human adaptation. In unpredictable environments, specialisations for particular conditions are less advantageous than structures and behaviours that enable adaptation to variable conditions in general. Humans, for better or worse, have evolved to live with uncertainty. Our ability to adapt to radical changes in our environment has been the key to our continued survival, and relocating to new regions on the Earth's surface has played a central role.

The most recent period of Earth's geological history – the last 2.6 million years – encompasses the entire period over which our genus *Homo* has existed. The current ice age, which, yes, technically we are still in, began at the start of this period. Its causes are not entirely understood, but contributing factors include changes in atmospheric composition like the concentrations of carbon dioxide and methane; the movement of tectonic plates, resulting in changes in the position of continents and oceans, which affect wind and ocean currents; sudden events like volcanic eruptions; and also the impact of relatively large meteorites.

In 1981 a research vessel discovered an iridium anomaly in sediment cores collected in the southern

Pacific Ocean. Iridium is one of the rarest elements in Earth's crust, and therefore large deposits of it are potential evidence of extraterrestrial material (as with the layer deposited by the asteroid that wiped out the dinosaurs). Further research of the ocean floor revealed that a large impact occurred around 2.5 million years ago, when an asteroid perhaps as much as 4 kilometres wide collided with Earth. The impact would not only have generated a tsunami hundreds of metres high, but may have also plunged the world into the current ice age. Although Earth was already cooling at this stage, the impact could have catalysed the cycle of glaciations that followed.

As the ice age set in, climatic changes in Africa caused grasslands to expand and forests to contract. This environmental shift corresponds with the emergence of the first members of the *Homo* genus, *Homo habilis*, about 2.4 million years ago. The genus *Homo* evolved from an Australopithecus species that had lived in forested areas for over a million years without appreciable evolutionary change; arboreal activity played a central role in providing places of refuge and food sources. With the shift in climate and landscape (and the snacking on psychedelic mushrooms blooming from dung as a supplementary food source if we are fans of the stoned ape theory), selection pressure for superior intelligence enabled the development of tool manufacture and advanced social behaviour, which in turn allowed for prolonged childhood development and the evolution of a larger brain.

Then, suddenly, from a million years ago, we see the rapid evolution from early *Homo* to us, *Homo sapiens*; this was an era of dramatic climate shifts, and as the only surviving *Homo* species we evolved characteristics to navigate such conditions.

Ever wondered why the weatherman so often gets it wrong? Climate on Earth is an extraordinarily complex phenomenon, making it even more challenging to predict weather in the short term than to understand longer-term climate shifts. The periodic changes in the environmental conditions on our planet can be caused by factors on Earth, but we also need to look beyond it. A simple example is the twenty-four-hour day–night cycle that takes place as our planet spins on its axis round the Sun. Our relationship with the Sun is naturally a primary factor driving climate, but it's more complex than just our rotation. Geophysicist Milutin Milanković described the collective effects of changes in Earth's movements on its climate on scales of up to millions of years, using three major contributing factors to these so-called Milanković cycles.

The first of these factors is the variable eccentricity – a measure of deviation from circularity – of our orbit around the Sun. The Earth orbits the Sun in an ellipse which is sometimes more elliptic than at other times. The eccentricity of our orbit varies on approximately 100,000-year cycles, primarily due to the gravitational pull of the massive planets Jupiter and Saturn in the

Outer Solar System. We are currently approaching our most circular orbit in the cycle, a time when the lengths of each of our seasons are about equal.

The second factor is the variation in the tilt of the Earth's axis of rotation. This tilt, with respect to the orbital plane as it travels around the Sun, or its obliquity, is the reason why Earth has seasons. The angle varies between around 22 and 25 degrees over a period of about 41,000 years; the greater the tilt, the more extreme are our seasons. Earth's axis is currently about halfway between its extremes, and will be slowly decreasing to its minimum over the next 10,000 years.

The third and final factor is the motion of the Earth's spin axis over time, circling around a direction perpendicular to the plane in which we orbit the Sun in a 26,000-year cycle. Also called precession, this wobble of the spin axis results in changes to the position of the Earth relative to the Sun at specific times in the seasons, making seasonal contrasts more extreme in one hemisphere and less extreme in the other. Currently, precession is making southern summers hotter and moderating seasonal variations in the north.

While our steadily orbiting Moon counteracts these disturbances, limiting for example the range of variation in obliquity, Milanković cycles, and their complex interactions, contribute to variations of up to 25 percent in the amount of incoming solar radiation at the Earth's surface; and a variety of studies have found that they show remarkable correspondence with periods of major climate change, in particular periods of glaciation or ice ages, over the past few million years. Beyond just the impact of the total incoming solar radiation, Milanković cycles indicate that a change in the timing of that heating can have drastic global results.

Another extraterrestrial factor that impacts climate on Earth is our Solar System's path through the Milky Way Galaxy; our Sun orbits the supermassive black hole at the galactic centre roughly every 230 million years. A study of zircon crystals from some of Earth's oldest remaining surface rocks indicates that the formation of our continental crust goes through cycles, with periods of increased crust production corresponding to times when we pass through major spiral arms of our Galaxy, which are also turning around the supermassive black hole at a slightly different rate. In addition, our Solar System undulates in and out of the galactic plane during its galactic traverse. Both of these movements entail our Solar System moving in and out of regions of space with dense interstellar clouds, resulting in more space matter crashing to the surface of the Earth, and in enhanced production of continental crust as well as environmental disturbances. The last time we encountered a spiral arm, according to one study, was around 66 million years ago – which would correspond with the extinction event that wiped out the dinosaurs. While our next encounter with a major arm is estimated to take place in over 100 million years' time, our next approach to the galactic plane may be sooner; still a comfortable 30 million years away.

Adding a further complication to the mix, palaeo-magnetic records tell us that the Earth's magnetic poles reverse rather regularly: in the past 20 million years we see a pattern of a pole reversal about every 250,000 years. The last reversal occurred around 780,000 years ago – which seems to indicate we are overdue for another – and since then the field has almost reversed fifteen times in so-called geomagnetic excursion events, where the field drops significantly in strength, not quite reaching the threshold needed to flip, before increasing again. The most recent excursion 42,000 years ago was brief, with the magnetic poles returning to their original positions after 400 years. During this excursion we think that our magnetic field may have weakened to just 5 percent of its strength today, resulting in increased levels of high-energy radiation reaching the Earth, causing a decrease in atmospheric ozone and changes in atmospheric cir-culation and climate, and possibly contributing to the extinction of the Neanderthals. *Homo neanderthalensis* specialised in hunting large mammals and lived in smaller groups on average and with a total population size less than *Homo sapiens*; this magnetic excursion and the resulting dramatic shifts in climate and food availability may have been contributing factors to their disappear-ance from the fossil record around 40,000 years ago.

As we can see, there are many factors at play that determine fluctuations and shifts in the Earth's climate. Milanković cycles can help us understand global trends in climate change, with implications both for current and future conditions on Earth, and corresponding cycles on Mars provide insights into the climate history there, both of which we'll return to later.

For now, back to the story of our origins: at the start of the ice age around 2.6 million years ago, climate cycles were dominated by the Earth's tilt for periods of 41,000 years. Then, around a million years ago, cold, dry periods became longer and more extreme, corresponding to a dominance (which has yet to be fully understood) of the 100,000-year cycles of variations in the eccentricity of the Earth's orbit. And out of this period of cold emerged our capability to control the most fundamental of resources, energy. Equipped with fire, the evolutionary enlargement of our ancient ancestors' brains relative to body size happened slowly at first, but then became especially pronounced over the past 800,000 years, coinciding with this period of more extreme climate fluctuation. Our large brains enabled improved storing, processing and sharing of information on novel environments within increasingly complex communities; our natural curiosity, creativity and inclination to collaborate were demonstrated to be successful characteristics to navigate great change. And more great change was indeed on the horizon.

The first *Homo sapiens* walked the Earth sometime after fire was first mastered, perhaps more than 200,000 years ago. In a cave on the southern coast of South Africa there is evidence of continuous human habitation between

164,000 and 35,000 years ago. During this time, humanity came breathtakingly close to extinction: there is genetic evidence indicating that the total population size of all of our ancestors plummeted to just 1,000 reproductive couples around 70,000 years ago. What could have been the cause of this sharp population decrease, and how did we survive?

About 74,000 years ago, a supervolcano on what is now the island of Sumatra erupted; an estimated 3,800 cubic kilometres of rock were blown out of the Earth, spewing billions of tonnes of ash, lava and gas into the air. Ash deposits found in Lake Malawi over 7,000 kilometres away indicate the scale of this event, and some idea of the terror that our (surviving) early ancestors must have felt as an explosion loud enough to be heard travelling multiple times around the Earth rang out, monstrous tsunamis ravaged coastlines across the planet, and the sky went dark with ash clouds that would have obscured the Sun for days, slowly falling to Earth and wiping out vast areas of vegetation even thousands of kilometres away.

The so-called Toba event is the largest known natural disaster to have occurred in the past couple of million years, with an explosivity dwarfing Krakatoa by two orders of magnitude, and a few times greater than the Yellowstone eruption of around 640,000 years ago. The resulting injection of noxious gases into the atmosphere and dust deposition had a global impact on weather and climate; the resulting volcanic winter is thought to have caused global mean cooling of perhaps as much as 15 degrees Celsius for a few years, as well as playing a role in the 1,000-year-long cooling period that followed. Pollen grains are often well preserved in layers of sediments and can become fossilised within rocks, providing a good source of data about past climates. Pollen analysis suggests that prolonged deforestation resulted in some parts of the globe, while genetic evidence indicates that several species of large mammals still roaming Earth today are all derived from very small populations around this same time.

Tiny glass shards produced during the eruption have also been detected on the southernmost coast of Africa. Thanks to food sources resilient to (even such dramatic) fluctuations in water and air temperature – in particular a year-round abundance of shellfish and a rich diversity of edible and medicinal plants, including subsurface roots and tubers – the few hundred kilometres of coast-line (including the Tsitsikamma region where we built our cabin) was a place where humans could not only survive during this time but also thrive, producing some of the earliest known symbolic art dated at 73,000 years old: lines drawn on a stone with an ochre crayon.

Climate change has been a frequent driver of migration. Communities developed new tools and ways of life to adapt to conditions in novel environments, and as a result we are adept at living in a range of different conditions, from the desert to the Arctic and everywhere in between, with a rich and diverse set of cultural practices to do so. *Homo sapiens* migrated out of Africa during times of unprecedented cold. Within the current ice age, there are glacial periods well predicted by Milanković cycles; the last glacial period extended from around 120,000 to 11,000 years ago. Compounded by the Toba super-eruption and resulting volcanic winter, a peak of cold, arid conditions, with sea levels at the lowest in a couple of million years, occurred around 70,000 years ago, at a time of minimal obliquity in the 41,000-year period of variation in Earth's axial tilt; low obliquity is favourable for ice ages due to ice sheet growth during periods of mild northern summers. This cooling was exacerbated by our being furthest from the Sun during the northern summer at this time. While some communities like those on the southern coast of Africa were thriving, for other, less-resourced regions these conditions would have driven many humans to seek new lands, in migrations that ultimately populated the entire world.

Aboriginal Australians are the oldest population of humans living outside Africa, descended from a single founding group that arrived in their new continent at least 65,000 years ago, bringing with them art and culture, seen in pieces of ochre used by artists dated to 60,000 years ago. Research into sea levels and ocean currents shows that travel to the supercontinent of then-joined Australia and New Guinea was likely achieved by purposeful and coordinated ocean voyaging, which would have required advanced cognitive, technological and cooperative capabilities. The Aboriginal people probably used dugout canoes to travel south of Asia; then, using visual cues, they followed precious freshwater and rapidly spread across the supercontinent, much of which was even more arid than it is today. Genetic studies of contemporary populations indicate minimal interaction with other cultures until the arrival of the British in 1788, by which time the Aboriginal population had reached around a million people. During this period of cultural isolation, stories were faithfully passed through generations for thousands of years, including chronologically accurate descriptions of sea-level rise as the last ice age subsided, animals that are now extinct, floods, asteroid strikes, volcanic eruptions and cyclones.

Our westerly migrations were likely triggered by both the resource scarcity and the extent of northern ice concurrent with the last glacial maximum beginning around 26,000 years ago, with sea levels more than 100 metres lower and average temperatures up to 5 degrees Celsius cooler than today. Mostly, the low obliquity at this time again resulted in cooler northern summers and the expansion of vast ice sheets over the northern regions of America and Europe, and the first groups of humans migrated across this frozen tundra to the Americas from Siberia no earlier than 23,000 years ago. Genetic analysis reveals a severe population bottleneck, indicating that the founding population of all native peoples of the Americas was just a few hundred, who, prior to the arrival of European explorers in 1492, had grown to a total population of around 60 million.

Since around 20,000 years ago, the Earth's orbit around the Sun has begun to approach circularity, with more balanced seasonal duration in the cycle of the Earth's orbital eccentricity. The ice sheets that buried much of Asia, Europe and North America stopped advancing, and high obliquity, coupled with a large, positive precessional parameter, contributed to a peak of incoming solar radiation during northern summers, the melting of ice sheets, and a rise in sea levels. But even this period of warming was punctuated by another global upheaval: pollen evidence shows that around 12,800 years ago there was a rapid cooling period when, in the space of just a couple of years, average temperatures dropped abruptly by as much as 7 degrees Celsius. This cooling period, called the Younger Dryas, put a range of species of large animals, including giant sloths, sabre-tooth cats, mastodons and mammoths, on the path to extinction. And what did it mean for us?

Francis Thackeray and his team discovered a platinum spike in peat deposits in South Africa from the same time, in addition to similar detections in over a dozen other sites around the world. This is compelling evidence for the theory that the Earth was struck by a platinum-rich asteroid, and that the global atmospheric dispersal of metallic dust caused the observed cooling. A sudden and powerful disruption of the Earth's crust, whether by impact or eruption, causes a rapid change in atmospheric conditions which can then result in dramatic global climate change. While the site of this potential impact remains to be confirmed, it would be the most recent natural disaster of a global nature that humanity has faced.

As things warmed up even more after the Younger Dryas cooling period, sea levels began approaching modern levels, towards a maximum last seen before the current glacial period spanning the previous 100,000 years. And these seas were the arena in which much of our exploration played out as modern civilisation was born in the relative calm of the past 12,000 years.

Enabled by our big brains, bipedalism, tools and cooperative social structures, over the past few hundred thousand years – driven by dramatic climate-changing events including galactic turbulence, geomagnetic flips, supervolcanic eruptions and asteroid impacts – we have expanded our society across the surface of our planet. As curious and collaborative creatives living in an often unpredictable environment, we have strived to understand and influence our environment and ourselves, culminating in rich and diverse cultures as well as rapidly developing technological capabilities. Just look at all we know about the Universe beyond the Earth; curiosity has motivated our study of the Cosmos and the worlds in it.

What can we learn from our ancestors who overcame extreme climate fluctuations in the past? Beyond their physical characteristics, they adapted by developing strategies to survive change. These included using tools to access resources and refuge, working in groups to advance skills and learning, as well as the ability to migrate long distances to new regions of the planet to establish new communities. From these strategies, not only for survival but for making sense of the world and our place in it, emerged culture. Ways of seeing the world and living in it were important enough in themselves to become drivers of migration; we embarked on journeys into the unknown to uphold ways of life, but also with the aspiration of creating new worlds as manifestations of our beliefs.

Seeking new worlds

Eugène Marais, an ancestor of mine and also a descendant of the original Marais family who arrived in what is now South Africa in 1688, was born in Pretoria in 1871. Eugène was an advocate, journalist and poet, and a pioneer in the scientific study of animal behaviour of creatures from termites to wild primates. His observations of evolutionary biology include: 'A species far-wandering and equipped for distant migrations, through inborn wanderlust or otherwise, would always have a better chance than one confined.' As humans, our innate curiosity and desire to explore beyond the world as we currently know it, together with our tendency to innovate and collaborate, enabled vast migrations across our planet that have both defined us culturally and ensured our survival in the face of dramatic changes to our environment.

We had all been nomadic for the vast majority of humanity's 200,000-year history; however, this was about to change. The temperate conditions that characterised the beginning of the Holocene era are associated with less pressure to migrate, and the establishment of our first permanent communities enabled further complex-ification of our societies. Jericho in Palestine is thought to be the oldest town in the world, with archaeological evidence of human occupation dating back 11,000 years. Permanent settlement saw the development of things like agriculture – a notable leap from controlling fire towards the large-scale shaping of our environment – trade and architecture, as well as increased opportunity for learning, and the emergence of social phenomena like organised religion and written records. It also gave rise to notions of nations and ownership that had not pre-viously existed, and within these larger, more complex societies the categorisation of people into hierarchical groups based on factors like ethnicity, gender, wealth and power occurred, coinciding with the emergence of slavery, or the ownership of one human by another.

The increased density of populations living in early villages tended to necessitate more resources than were available locally, and the development of transportation systems like ships enabled trade, resource acquisition and continuous exploration. While there are claims that

Chinese explorers were travelling and interacting with local people all over the world, including the Americas, over 2,500 years ago, the earliest documented intercontinental voyage was led by explorer and diplomat Zheng He in the early 1400s. In 1405 he departed from China, with a fleet of 317 ships and around 28,000 crew. These fleets visited South East Asia, India, the Horn of Africa and Arabia, trading goods and improving the accuracy of global mapping along the way. By contrast, Portuguese explorer Vasco da Gama's fleet in 1498 contained four small ships and at most 170 men.

Following the death of the Yongle emperor, who had supported exploration and expansion, in 1424, and after the last 'Treasure Fleet' in 1433, China began to close itself off from the outside world. Subsequent Ming emperors dismantled their great ships and invested the large amounts of money previously devoted to maritime exploration into building the Great Wall and maintaining huge armies to ward off the Mongols on the northern borders. This was a time when Europe became more powerful.

European powers such as the British and Spanish empires emerged, launching grand journeys into the 'unknown' and more often than not finding people living there already. The (typically genocidal) establishment of remote colonies in Africa, the Americas, Asia and Australia ensured a continuous extraction of resources to increase power and wealth back home. These 'resources' included humans, and formalised slavery, in particular the transatlantic slave trade, escalated. At the same time, the emergence of organised religion fostered a sense of community through shared beliefs between people who may never have met, while also enabling the remote control of societies extending over vast geographic regions.

While permanently inhabited regions became a feature of human life, this did not put an end to our movement around the planet. At the same time as the slave trade saw people forcibly moved over intercontinental distances, doctrine-driven migration emerged as a new social phenomenon. Motivated by more than the pursuit of resources, communities moved great distances to be able to practise their particular cultures, and in the new conditions, often integrated with other communities already there, new ways of thinking and organising society emerged.

My Marais ancestors set sail for the Cape of Good Hope in December 1687 on the *Voorschoten*, the ship carrying the first French Huguenots from Europe to the southern tip of Africa. They were Protestant peasant farmers escaping drought and poverty, as well as unaffordable taxes and religious persecution by the Catholic authorities in France. They had very little idea of where they were going, not many people having been to and returned from the newly established Dutch resupply colony at the Cape. Months after their departure, they landed 140 kilometres north of their intended destination, at Saldanha Bay, due to damage caused to the ship by a severe storm. A messenger was sent on foot to Table Bay to inform the Cape governor, who sent a smaller ship to fetch the immigrants, who arrived at the fort of what is now Cape Town in April 1688.

All members of the Marais family – Charles, his wife Catherine Taboreaux and their four children – survived the journey, and settled in a valley almost 100 kilometres from the fort. The initial years of settlement were tough, with a land and climate different to what they knew. In spite of the challenges, they established irrigation systems, mills, crops, orchards and animal stock from scratch. Just one year after arriving, Charles died of injuries sustained in a 'dispute with a local'. Catherine and her children went on to found what is now Plaisir de Merle Wine Estate in Franschhoek outside Cape Town, as well as lay the foundations for a prolific population of Marais descendants in South Africa. Regretfully, my family no longer participates in the vineyard, apart from enjoying some of South Africa's best wine, in particular wine-maker Niel Bester's flagship Charles Marais Bordeaux red blend (which is superb). The French Huguenots went on to play an important role in the development of the nation that is now South Africa, in particular to its legislation and economics, as well as Afrikaans culture and wine-making. This is just one story of the grit and resilience that have enabled human communities to relocate vast distances across the surface of our planet, driven by belief in their way of understanding the world into what was to them the complete unknown, and writing new chapters in human history as the result of their bold venturing.

Another story is also about a boat that travelled across an ocean with religious refugees on board. The *Mayflower* set sail for the 'New World' west of Europe over a month behind schedule from Plymouth in September 1620, with around 130 people on board. Provisions were already running low due to the delay, some of the passengers having been on board since their departure from London almost two months before. The crew of around thirty included the captain, mates and seamen, a surgeon, a carpenter, stores managers, fishermen, cooks and gunners. Just over a third of the 100 passengers were Pilgrims, seeking religious freedom from persecution by the Church of England, and included three pregnant women and others who were adventurers, farmers, hired hands or servants recruited by London merchants to work at the Colony of Virginia.

Their cargo comprised stores for the journey and their new lives, including food, building materials, cannon and gunpowder, dogs, sheep, goats and poultry, and also two boats. They had a compass for navigation, an hourglass to measure time and a log and line system to determine speed in knots. Huge waves during the Atlantic crossing fractured a key structural support timber on their ship, which was repaired at sea with equipment brought to construct the settlers' homes.

In November they sighted land, and despite attempts

to travel south to their planned destination of the Colony of Virginia, strong winter seas and dwindling supplies forced them to anchor at what is now Provincetown harbour on the Cape Cod peninsula. After dropping anchor, forty-one of the male passengers signed the Mayflower Compact, a social contract in which they consented to follow their own set of rules and regulations for the sake of order and survival; the first governing document of what became the Plymouth Colony. Apart from trips to shore, when the men took corn and grain from buried native stores and allegedly raided burial sites, for the duration of the bitter winter they lived on board the ship, suffering outbreaks of contagious diseases from the lack of fresh air below deck.

Taking refuge in a bookstore one stormy afternoon in Provincetown on my way to a conference in Boston, I was intrigued to hear some details of the story from a member of the Pilgrim Memorial Association. She told me about the women on the trip. One woman's baby was stillborn while still docked in Plymouth, one woman gave birth at sea, and one while anchored at Cape Cod. While eighteen women boarded the ship in England, only five made it through the first winter; many died before they were able to set foot on land because they stayed on board, often below deck, to care for the young and sick in damp, cold, confined and unsanitary conditions.

The following spring, the survivors disembarked to build huts ashore. Only just over half of the original 130 people had survived. While, notably, those for whom

the freedoms and liberties were proclaimed in the Mayflower Compact excluded women, most of the crew and
all the servants, the now-famous document was the first
to declare self-governance in the 'New World', arguably
evolving into the constitution of what is now the United
States of America. Some attribute the emergence of
democracy in the US, and perhaps at the same time the
still-unresolved inequality built into the system, to this
small boat of pioneers.

We have evolved to inhabit this planet. We can breathe the
atmosphere, drink the water flowing across the surface and
eat the plants and other species with which we share the
environment. Leaving the planet is to leave 4 billion years
of acclimatisation behind, and enter environments where
the resources we rely on are not so readily available. However, parts of Earth can also be pretty extreme; meaning
that things like temperatures, availability of water, oxygen
or food are beyond the threshold of what much of terrestrial life on Earth, in particular humans, finds comfortable
or indeed survivable. Increasingly, we don't even need to
travel to remote places to experience this extremity: the
prolonged droughts, heatwaves, wildfires, desertification,
flooding, storms and periods of extreme cold predicted to
occur as a result of our impact on our natural environment
are already affecting the places where we live.

The silver lining to all this is that these are great conditions under which to prepare for life off-world while
simultaneously developing solutions for resilience in the

face of climate change on Earth. Today we are equipped with technologies that have advanced significantly over the past million years since mastering fire. However, if we choose to embrace them, also at our disposal are the characteristics that make us human: the capabilities that enabled the survival of early humans in spite of their rudimentary equipment compared with what we are used to today. The solutions for our future resilience cannot rely on technology alone: individual and community adaptations play a make-or-break role, as those who have been part of a team in an extreme environment, in particular when technology fails, will know.

Out in the cold

'Men wanted for hazardous journey. Low wages, bitter cold, long hours of complete darkness. Safe return doubtful. Honour and recognition in event of success.'

In 1912, Earnest Shackleton, allegedly anyway (there's a 100-US-dollar reward out for a hard copy), published the above advertisement in a London newspaper. Whether actually published or not, strong parallels are apparent between the volunteer expeditions to the final frontier of the early twentieth century, the continent of Antarctica, and the imminent first crewed missions beyond our Earth–Moon system.

One parallel that we'll be doing away with, however,

is the idea that it will just be men who will embark on the hazardous journey. When you stop to think about it, it's clear that women are far better suited for space exploration; some may say it's because they are more intelligent. But seriously, not only are women on average smaller and therefore weigh less than men, but they also breathe less, drink less and eat less. This is good news for mass-restricted space missions. Coupled with our knowledge that diverse teams are better at solving unforeseen challenges, it is a no-brainer that, just like in *Star Trek*, gender-equal teams with a range of cultural backgrounds and skills are best suited for navigating new environments.

Other than that, Shackleton's advertisement has remarkable parallels with the headline that I read in a newspaper exactly a century later, in 2012: 'Call for volunteers for one-way trip to Mars'. And now is as good a time as any to digress briefly on the Mars One Project and the people who volunteered to go out beyond Earth to the cold conditions of the planet next door and stay there forever.

The Mars One Project never got off-world, but it got a lot of people, including me, excited about our imminent expansion beyond Earth. I clearly remember my reaction to reading the headline, which was firstly one of nausea when I knew that I would of course devote myself entirely to being on this trip. Secondly, it was one of excitement, as I realised that the 'one-way' aspect would make the application process a lot less competitive. In 2015, I was over the moon to be shortlisted along with ninety-nine others from around the world as an astronaut candidate with Mars One.

However, putting the first humans on Mars for good is a hugely ambitious proposal by a private company, a start-up moreover, without any government affiliation or billionaire founders, and I don't think that anyone should be too surprised that it didn't pan out as planned. For those of us who were involved since 2012, our journey with the project was rarely without controversy. Mars One faced fundraising challenges since the outset, had a fatwa declared against it in 2014, and finally declared bankruptcy in 2019. The mission to Mars, however, is bigger than any one of us. And we'll get there. One way or another.

The Mars One Project did enjoy great successes here on Earth, however, highlights of which include being featured on some of my all-time favourite TV shows: Sheldon Cooper volunteers to go to Mars with the Mars One Project on *The Big Bang Theory*, Cartman volunteers on *South Park*, as well as Lisa on *The Simpsons*. The project succeeded in making an impact on popular culture as well as getting people outside of the space industry thinking about human exploration beyond Earth. In many ways, Mars One lit the fires of excitement that moved the idea of space settlements from a realm of pure science fiction into one of possible reality. But before we get to Mars, let's return to a region on Earth that provides a great analogue for an off-world environment.

As the Earth spins on its axis while orbiting the Sun, the Poles receive the least sunlight; these are the coldest regions on the planet. While the Arctic is an ocean surrounded by land, Antarctica is a continent with elevation surrounded by ocean, making it the coldest, driest, windiest, highest and most isolated place on Earth. In the interior of Antarctica, the Russian base of Vostok is almost 3,500 metres above sea level, and in 1983 the lowest temperature ever recorded was measured here: almost negative 90 degrees Celsius. Wind speeds in Antarctica have been registered at over 300 kilometres per hour.

Antarctica is arguably more remote than Earth orbit or the Moon; during the winter there is no way in or out; no boat can crack through the sea ice to access the continent and few weather opportunities present themselves for brave pilots to risk the dark, windy and treacherous conditions that prevail. With average surface temperatures on the Red Planet of around negative 60 degrees Celsius, Antarctica, particularly through the winter, provides the closest all-round analogue on Earth to living on Mars.

In the Arctic, commercial traffic now travels along once-treacherous routes that have become passable due to global warming and the melting of Arctic Sea ice at a rate of nearly 13 percent per decade. Antarctica, on the other hand, remains (mostly) free of commercial activity and almost completely inaccessible for more than six months of the year, during which time around 1,000 inhabitants maintain research activities at several dozen permanent national stations around the continent. When

I arrived there in December 2019 after a five-hour flight from Cape Town I thought to myself, 'This is the closest I have been to experiencing what it will be like to explore beyond Earth' – only to discover that I knew several people working at the base there, including one of our hosts whose father was the principal at my brother and sister's school in Pietermaritzburg. Small world indeed. Or maybe people from 'the Pit', as I like to call it, are just trying to get as far away from home as possible!

While today overwintering researchers typically enjoy heating and Internet and other comforts of modern life while operating state-of-the-art scientific equipment installed in their national bases, things were different a century ago.

Arctic exploration by ship goes back thousands of years. But only in 1906 did Norwegian explorer Roald Amundsen and his crew of six successfully navigate the so-called Northwest Passage, a much-sought-after sea route between the Atlantic and Pacific Oceans through the Arctic Ocean rather than all the way round the continent of South America. During this voyage they spent almost two years living on an island, interacting with the local Inuit people, learning to make clothing, participating in dog-sledding and seal-hunting activities in the winter and kayaking and fishing during the summer, in addition to doing magnetic observations in their search for the North Pole.

The Inuit have lived in the Arctic for at least 1,000 years, following migrations from north-eastern Siberia a few thousand years before that. Historically, they lived in nomadic family groups governed by elders that joined together in bigger groups in winter, when they would meet at hunting spots for the free exchange of food, supplies and news. Traditionally, Inuit people ate mostly meat, the primary biomass available in Arctic conditions, and analysis has found unique genetic mutations for adaptation to cold as well as a diet high in omega-3 fatty acids, which also results in reduced stature. Along with physical adaptations to the climate, in Inuit culture the land, consisting of the earth, water, ice, wind, sky, plants and animals, is sacred and part of an interconnected wholeness.

The experience of living with the Inuit was to prove invaluable to Amundsen in the pursuit of the South Pole five years later. He had secured funding for a trip to the North Pole in 1911, but believing he had been beaten to it (in what turned out be an inaccurate report), he decided to go south instead. He didn't tell any of the crew until they reached Madeira off the north-west coast of Africa, and then navigated south instead of north: towards Antarctica. (Amundsen did in fact get to the North Pole by plane in 1926; it was only reached by land in 1968, just a year before the first Moon landing.)

Besides the mysterious Piri Reis world map compiled in 1513 from older maps that appears to show details of the coastline and interior of Antarctica at a time when it was largely ice-free, which would have been at least 6,000 years ago, the southernmost region of the planet remained mostly mysterious until a Russian expedition in 1820. It wasn't clear at that time if Antarctica was a group of islands or a continent, and the interior was only accessed early the next century.

Led by British sea captain Robert Falcon Scott, with Irish master mariner Ernest Shackleton on board, the *Discovery* voyage to Antarctica in 1901–4 was a prolific scientific mission that also made significant progress in mapping the coast and interior. Shackleton's subsequent unsuccessful attempt in 1907 to reach the Pole involved dogs, ponies and a purpose-built motor car. However, it was heavy, with little traction, and its petrol engine performed poorly from the outset. Shackleton and his team did, however, get within 100 miles of the South Pole.

Once Amundsen's mission of 1911 arrived in Antarctica, his Inuit-based strategy for travel involving only packs of dogs proved to be more efficient. His small team made use of traditional Inuit clothing made from animal skins and included experienced dog drivers and master skiers who could keep up with the sled dogs. Seals and penguins provided essential calories and nutrition for both humans and dogs, as they had learned in the Arctic from the Inuits' predominantly meat diet. He planned for the weaker animals to be eaten, departing with fifty-two dogs and coming back with eleven. They succeeded in reaching the Pole in December 1911 and were back on the ship within ninety-nine days. They stopped at Hobart, Tasmania, on the way home to send a telegram back to Norway stating simply: 'Pole reached'.

When attending an event in Hobart I visited this historic post office, where a plaque commemorates the message.

Scott set out just weeks after Amundsen. He planned to use ponies for the first quarter of the total journey and man-haul with dogs for the rest. He also brought a kind of snow tractor, which broke down almost immediately, but there was no mechanic to repair it. Scott's pony fodder had to be brought all the way from England in their ship. Both Scott's and Amundsen's men had around 4,500 calories of food per day, but while Scott's team were pulling their sleds on foot Amundsen's men were skiing next to the dog sleighs, expending far less energy.

Scott's expedition objectives were twofold: to reach the Pole and to explore and document this great southern land. Scott and his team reached the Pole just five weeks after Amundsen, but did not survive the journey back. In the tent alongside their frozen bodies were 16 kilograms of fossils, a meteorological log, scores of notes and rolls of film; additions to knowledge in many branches of science. These first-ever found Antarctic fossils included an example of a now-extinct beech-like tree dated back to 250 million years ago, fossils of which were also discovered in India, Australia and later in Southern Africa and South America, confirming that Antarctica had once been part of the ancient supercontinent Gondwanaland, with a climate mild enough to support trees. It was also the first time a film camera had been used to make breakthroughs in the study of biology, setting the standard for future expeditions as well as wildlife documentaries.

Scott's final diary entry included: 'Had we lived, I should have had a tale to tell of the hardihood, endurance and courage of my companions which would have stirred the heart of every Englishman. These rough notes and our dead bodies must tell the tale.' The Norwegian explorer Tryggve Gran, part of the search party that set off to find the missing Scott and his team in 1912, said: 'When I saw those three poor souls the other day, I envied them. They died having done something great.'

After losing out on being the first to reach the Pole, Shackleton's subsequent mission in 1914 was an attempt to cross the continent for the first time. Shackleton's ship, the *Endurance*, however, did not make it to the continent of Antarctica. It was frozen into the Weddell Sea during an unexpected cold spell during the summer of their arrival in the region. The ship was converted to a base where they overwintered. However, by the next spring the ship began breaking up under the pressure of the ice, and the crew had to evacuate with lifeboats and some supplies, and then watch as their home and means to get back to civilisation creaked and cracked and finally sank beneath the ice. 'I cannot write about it,' said Shackleton in his diary.

With their ship sunk, Shackleton and his team of twenty-seven men made camp on an ice floe. They drifted northwards until the floe broke up, now having to contend with the open sea in lifeboats not fit for purpose. Captain Frank Worsley navigated for six days at sea, not sleeping for eighty hours, while the crew suffered sea-sickness, dysentery and bitterly cold and wet conditions. Shackleton's second-in-command, Frank Wild, wrote that 'at least half the party were insane' by the time they miraculously arrived at one of the only land masses in the entire region of vast open sea: the hostile and uninhabited Elephant Island. From there, Shackleton and five men sailed over 1,300 kilometres in a small boat to South Georgia to get help for the rest who stayed behind. In another feat of extraordinary navigation by Worsley, they made it. It took another four months before Shackleton was able to rescue the stranded men on Elephant Island, but he succeeded, with not a single life lost. Three of the crew, however, returned only to fight and die in the First World War, which had broken out while they were lost at sea for two years.

In 2022, a group of scientists and explorers headed to the treacherous waters of the Weddell Sea to search for the *Endurance*. They sailed on the South African icebreaker and research ship *SA Agulhas II*, equipped with helicopters and remote-controlled underwater vehicles. They found the legendary ship just a few kilometres from where Worsley recorded it to have been lost beneath the ice. The ship is in remarkable condition on the ocean floor 3,000 metres below the surface, its name clearly visible on the stern.

Polar travel has come a long way since the *Endurance* was above water: explorers Ben Saunders and Tarka L'Herpiniere were the first to journey to the South Pole (and back again) on foot via Scott's same route over a century later, in 2013. Ben and Tarka's cargo included freeze-proof laptops, freeze-dried meals, electrolyte drinks and a mobile satellite hub. They consumed almost 6,000 calories a day, while burning more than 9,000. They covered around sixty-nine marathons back to back dragging 200-kilogram sleds of supplies each. Saunders lost 22 kilos during the expedition, and when I met him in Cape Town on his way back from another trip to Antarctica he told me that it took him a long time to stop psychological feelings of hunger. 'We put ourselves through hell,' he said.

Of all the personalities and strategies involved, Amundsen got to the South Pole first, with a small team entirely focused on this objective. Crucially, his experience and knowledge gained with the Inuit was invaluable in his mission design; his strategy was based on a synergy with the environment that included utilising local food sources like seals and penguins, the skins of animals evolved for polar conditions for warmth, as well as a team of people and dogs with the mastery to move swiftly through the treacherous terrain.

During the so-called Heroic Age of Antarctic Exploration, everyone had more than one designation: engineers became pioneers of Antarctic photography and film-making, while medical doctors painted watercolour landscapes. Pets including a cat and a rabbit made it down to Antarctica too, besides the pack dogs and ponies that played a central role in everyday life. Along-side the physical heroism are the art and journal entries indicating the horrors experienced, but also the strong sense of community spirit and wonder at the inexorable beauty of the landscape that emerged among the men. Contemporary Inuit have had difficulty adjusting from living free in such an otherworldly (to non-polar inhabitants anyway) landscape in the far north to their loss of self-sufficiency and self-determination within the confines of more urbanised life. Inuit suicide rates are among the highest in the world. What does it say about our society's potential to weather extreme conditions that people evolved to thrive in harmony with such harsh natural environments find it painful to adapt to modern life?

What a peculiar effusion of sentiments the welcome face of the sun draws from our frozen fountains of life! How that great golden ball of cold fire incites the spirit to expressions of joy and gratitude! How it sets the tongue to pleasurable utterances, and the vocal chords to music! The sun is, indeed, the father of everything terrestrial. We have suddenly found a tonic in the air, an inspiration in the scenic splendours of the sea of ice, and a cheerfulness in each other's companionship which make the death-dealing depression of the night a thing of the past.

So wrote surgeon Frederick Cook in the spring of the

first overwinter mission of 1897–9 to Antarctica. Their ship the *Belgica* was trapped in the pack ice for over a year, and Cook saved many lives, anticipating established medical science by decades, with his techniques.

Eighteenth-century philosopher Edmund Burke described 'the sublime' as the most powerful aesthetic experience of which the mind is capable, a mixture of fear and excitement, terror and awe. A number of authors have associated early polar exploration with the experience of the sublime, which has stronger parallels with the establishment of the first off-world bases than do our past few decades of technical achievements in Earth orbit. Antarctica remains the ultimate environment on Earth in which to test the equipment, psychology and team spirit required to venture into the harsh unknown beyond civilisation. It's a stark reminder, more than 100 years later, that we are far more resilient than we may imagine in this comfort-seeking modern era. Exploration drives innovation, both technological but crucially also cultural: sometimes the toughest of environments bring out the best in us.

In spite of all the challenges, extreme environments are places where the vast majority of teams complete their assignments successfully and harmoniously. For example, there has not been a recorded murder in any space station, or ever (officially anyway) on the continent of Antarctica. What may be an urban legend tells of two scientists playing a game of chess at the Vostok Station in 1959, the loser allegedly attacking the other with an ice axe, prompting chess games to be banned at Russian Antarctic stations. However, this tale remains uncon-firmed. In 2018, an attempted murder was reported: a scientist allegedly stabbed a colleague for giving away the endings of books. Considering the bleak and con-fined conditions under which thousands of scientists have been overwintering in Antarctica over the decades, I'd say this is not a bad track record. We *can* endure extreme environments; and determination, community and a strategy of collaboration with the environment and within the team are critical to doing so.

Into thin air

At 8,849 metres Mount Everest is the highest that the sur-face of the Earth reaches into space; its upper slopes are among the harshest environments on the planet. When asked by a reporter why he wanted to climb the moun-tain, George Mallory famously replied: 'Because it's there.' Mallory died in 1924 while attempting the ascent. Edmund Hillary and Tenzing Norgay first reached the summit of Everest in 1953. Norgay wrote in his autobi-ography that he was a Sherpa, a people indigenous to the mountainous regions of the Himalayas, while Hillary, a New Zealander, was described as 'exceptionally strong and possessed of a thrusting mind which swept away all unproven obstacles'. Hillary described Norgay as some-one of great personal ambition and physical confidence, with an irresistible flashing smile.

While thousands of people have summited Everest, just a couple of hundred of them have done so without supplemental oxygen. At 5,800 metres conditions are already challenging, with half of the oxygen available at sea level – like having a collapsed lung, I joked breathlessly approaching the summit of Kilimanjaro.

The highest permanent settlement in the world is at about 5,000 metres above sea level: the town of La Rinconada in Peru. Genetic studies of populations living permanently at high altitudes, defined as being anything over 2,500 metres, show the prevalence of several genes that provide protection from high-altitude hypoxia, or lack of oxygen, particularly in the regulatory systems of oxygen respiration and blood circulation. For example, studies show that many Tibetans make efficient use of these lower oxygen levels through the regulation of the body's production of haemoglobin, which facilitates the transport of oxygen in red blood cells, and it has also been documented that the children of Tibetan women with these genes for high blood-oxygen content had higher survival rates than the children of those who did not.

Neurosurgeon and climber Douglas Fields once proclaimed: 'Three attributes of a good mountaineer are high pain threshold, bad memory ... and I forget the third.' The first stage of high-altitude sickness is called acute mountain sickness, which can cause headache, insomnia, dizziness, fatigue, nausea and vomiting. The

next stage is high-altitude brain swelling, which is potentially fatal. Both are caused by the body's reaction to
low levels of oxygen, which eventually directly impairs
or damages brain cells. Understanding how to prevent,
anticipate and mitigate these kinds of reactions is critical to our exploration of low-oxygen environments.

Orthopaedic microsurgeon Ken Kamler is also an
experienced explorer, and was the only doctor on the
slopes of Mount Everest during one of the worst disasters there in history. He tells the tale of how one of
the most unforgiving environments on the planet is also
a place to see the power of the human mind in sum-
moning the will to live against all odds. In a period of
calm around midnight in May 1996, thirty-three climbers headed up to the summit from Camp IV on Everest,
which is at almost 8,000 metres. The next day, the wind
picked up with a ferocity that Kamler had never seen
in his previous expeditions on the mountain. In the following hours, as some of the climbers who had emerged
from the storm struggled back down the mountain suffering from snow blindness, frostbite, hypothermia or all
of the above, Kamler tended to them, as best as he could
at an altitude of 6,300 metres, while other climbers went
up to attempt a rescue.

Eight people died that day. But one man, who according to all medical understanding should have died, didn't.
Pathologist Beck Weathers had been reported dead after
lying in the snow at an altitude above 8,000 metres for
fifteen hours. Then, suddenly, he decided that he didn't

want to die. He stood up, made his way to Camp IV and eventually into Kamler's tent, where, completely lucid, he asked where he should sit and if his health insurance was accepted there. While suffering severe frostbite on his hands, feet and face, by the will to live he reversed what is thought to be irreversible hypothermia and made it home. We cannot underestimate the power of the mind when the body is pushed to its limits – something a culture preparing for extreme conditions should take into account.

Kamler's exploration has also taken him to another set of extreme conditions: under the ocean, to the Aquarius Reef Base situated 10 kilometres off the coast of the Florida Keys in the US, at a depth of 19 metres below the surface. The only facility of its kind, it was built as an analogue to the ISS and is where astronauts train before departing for Earth orbit.

Under the ocean is one place to experience life inside a habitat with an unbreathable but accessible environment outside. While an astronaut is classified as anybody who has been more than 80 kilometres above the Earth's surface (though, memorably, Jeff Bezos and his crew were not classified as astronauts according to the US Federal Aviation Administration, despite having achieved this altitude, due to fact that they did not 'demonstrate activities during flight that were essential to public safety, or contributed to human spaceflight safety'), an aquanaut is anyone is who stays at depth under the ocean for twenty-four hours or more. The first such person achieved this title in 1962. Underwater archaeologist Robert Sténuit spent twenty-four hours inside a tiny one-man cylinder at a depth of 61 metres on the French Riviera, but a storm surge caused the cylinder to float up to the surface, where Sténuit remained safe from decompression sickness because the cylinder was still pressurised. He then decompressed safely, becoming the world's first aquanaut.

Under the ocean, the pressure increases by around an atmosphere for every 10 metres of depth. At these increased pressures, the nitrogen gas in the lungs is compressed and can enter the blood vessels. If a diver returns to the surface too quickly, the nitrogen gas expands and can form bubbles in the blood and tissues. Decompression sickness is a potentially fatal condition, preventable by limiting a diver's rate of ascent to reduce the concentration of dissolved gases in their body.

Aquarius consists of three compartments: access to the water for scuba excursions is made via the 'wet porch', which is at the same air pressure as the water pressure at the base's depth of 19 metres; a small airlock compartment; and the main compartment where pressure can be varied. Living at depth for more than twenty-four hours results in dissolved gases in the body reaching saturation point. For any duration beyond this period, a resurfacing rate of around an hour per metre is required. Before returning to the surface, the aquanauts stay inside the main compartment for seventeen hours as the pressure is slowly reduced to sea-level pressure, to prevent decompression sickness. A life-support buoy on the surface houses power generators, a solar array and data connections, as well as water tanks and air compressors to replenish fresh air and water.

In 2022 I visited Aquarius director Tom Potts, who has been with the programme since its inception in the 1980s. I wanted to know more about what the limiting factors were to staying down there, other than saturation. Potts explained that the humidity, inevitable due to breathing and diving excursions, builds up to 100 percent inside the habitat. Also, the shore crew constantly monitoring life-support systems and aquanaut health need to be rotated. This sounded manageable, and I had immediate thoughts about breaking the current world record of 100 days underwater. Notably, however, the saturation diver and biomedical engineer Joseph Dituri, who set the current record in 2023, had been living alone in a nearby luxury facility rather than in a 37-square-metre living and working space with five other people built in the 1980s.

Habitats providing shelter from otherwise deadly environments are often space-limited, and living in close quarters with others is a challenge experienced by the vast majority of people in extreme kinds of environments. Whether it is the close quarters of an igloo, camping on Everest, an underwater room, a spacecraft or an off-world base, there are a number of practical challenges that need to be thought out, planned for and solved ahead of any kind of crewed mission.

I recently met an astronaut at an event celebrating

the first all-private crew being launched to the ISS by SpaceX, and we got to talking about Aquarius as he had spent some time training there and I had recently visited the facility. Toilets are one such practical challenge in these kinds of environments. He explained to me how the situation worked: being a confined space, the indoor toilet is only used for defecating during decompression. The rest of the time, aquanauts make ocean excursions to the 'gazebo' – a bell-shaped structure with a pocket of air in it – a few metres from the base. The primary function of the gazebo is as an emergency shelter, but it also functions as an outhouse, where various sea creatures eagerly clean up any emissions. Although I have done a fair amount of night diving, his description of solo gazebo missions in the dark was an eye-opener … to think of the sea creatures lurking there waiting for a midnight snack!

He also said, having spent time in the ISS, that living in the base is the closest experience he has had to being in space. There are many analogies: the need to suit up to explore the region beyond the habitat; the experience of zero gravity while doing 'extravehicular activities' (namely scuba-diving); and the emergency evacuation, taking seventeen hours for decompression (evacuation from the ISS can be even quicker than from underwater), making it one of the closest comparisons to living in Earth orbit. The conditions of isolation and confinement within a habitat providing necessary life support in an environment lacking breathable air also have parallels with the long-duration spaceflight necessary to travel to, and settle on, Mars – or beyond; and even more striking parallels with the exploration of the subsurface oceans of several Outer System moons.

Invaluable off-world, air may be the most underrated resource here on Earth; however, as our atmosphere warms up, we're beginning to pay more attention.

Feeling the heat

The year 2023 was the warmest year on Earth since global records began in 1850. And with the northern summer of 2024 the hottest on record, narrowly exceeding the record set in 2023, this is no passing fad. With pre-industrial (1850–1900) average temperatures – around 13.3 degrees Celsius – as a baseline, the twentieth century saw a rise of around 0.6 degrees, while in 2023 the average temperature rose to 15 degrees Celsius. While this may sound slightly on the cold side, for an African anyway, by some counts we are already beyond the 1.5-degree threshold agreed to at the UN Climate Change Conference in 2015. At this rate, some predict a 2-degree increase by 2030, taking our climate outside the range of the last several hundred thousand years, which will result in an increase in heatwaves, droughts, wildfires and potentially also volcanism, a significant drop in food production and availability of potable water, massive social disruption and, knowing us, further conflict as a result of scarcity of resources.

And yet we've managed to do the seemingly impossible and put a lander on Venus, which has surface temperatures of nearly 500 degrees Celsius; high temperatures alone do not have to spell the end of humanity. Indeed, we developed strategies without modern technology that prevailed for thousands of years for living in a range of hot, dry regions across the planet; we could learn a thing or two from these places.

The hottest temperatures ever recorded on Earth, so far, are in the 50s. Without sufficient shade and water, humans can perish in just hours in such conditions. When I visited Death Valley in the Mojave Desert in the summer of 2018, I stopped at the last service station for almost 100 miles, and the memorable character at the store, worried that I was travelling solo, told me, in a strong Californian accent: 'That's a lonely road south of Badwater. Something happens, you just don't know how long you'll wait.' The scorching sting to the nostrils and eyes of the hot air blowing off the saltpans at Badwater, nearly 90 metres below sea level, can only be borne for a short while. As I drove south, I watched the external temperature gauge rise to over 125 degrees Fahrenheit, and with no phone service to check the conversion I confirmed later that evening that temperatures had risen to 52 degrees Celsius. Thankfully, my rental's V6 engine stayed below half temperature the whole way, even with the air conditioning on: remembering the words of the man at the store, a feat of engineering I did not take for granted that day!

The hottest places on Earth are also typically dry, the lack of vegetation and water and abundance of sand giving rise to extreme temperatures. Desert sand consists primarily of silicon dioxide grains, which heat up fast in sunlight. The highest recorded temperatures on Earth are in such deserts. Under standard measuring conditions, all continents apart from Antarctica, Europe and South America (this may change in coming years) have measured temperatures over 50 degrees Celsius. While historical temperature readings can be controversial, according to the World Meteorological Organization the highest temperature ever recorded was nearly 57 degrees Celsius in 1913 in Death Valley.

Perhaps the most extreme hot location on our planet is the Danakil Depression in Ethiopia: 91 metres below sea level, it is on average the hottest place on Earth, not only during the day with the Sun beating down from above, but also at night owing to geothermal heat coming from below. There are two active volcanoes in the region, several crater lakes of lava bubbling up, and sulphuric hot springs. These environments are useful for the study of the microbial extremophiles that live there, in particular in thinking about how life might arise in similar extreme conditions beyond Earth. The bacteria found living in the water inside one of Dallol's volcanoes in 2019, at a temperature of 89 degrees Celsius, a highly acidic pH of 0.25 and surrounding air on average at 38 degrees, could give insights into potential life on highly geologically active early Mars.

This very otherworldly landscape is also referred to as the cradle of humanity; in 1974 Lucy was found there, an *Australopithecus afarensis* female – one of our oldest ancestors – dated at 3.2 million years old. More recently, the Afar people have lived in the Danakil Desert in Ethiopia for at least 2,000 years. While resources in the Afar region are scarce, salt-mining is a practice extending over many generations; in the hottest place on Earth, Afar men wielding only basic hand-held tools produce almost all of Ethiopia's salt. Although since 2020, when the conflict in the north of Ethiopia broke out, hundreds of thousands of Afar have been displaced from their land, they remain a proud and resilient people.

Conditions in Death Valley or the Danakil Depression may seem extreme to most modern humans accustomed to the comfort of air conditioning in even far milder conditions; however, people have lived in desert environments for millennia all over the planet, and more recently people living by ancient cultures have been forced to migrate there to continue their way of life. In the biggest known lava tube in Saudi Arabia, where conditions are much cooler than at the surface, evidence of human habitation may date back as much as 10,000 years. As we plan for an increasingly extreme future, both on Earth and beyond (we'll revisit lava tubes again later), we would do well to understand human resilience from the perspective of some of the longest-prevailing cultures on the planet.

The Namib is the oldest and one of the driest deserts on Earth. Like Mars, the sand has a distinctive rust colour from the iron oxide content and geological features that tell the tale of ancient water flows – until 55 million years ago, the desert was an ocean floor. Today, water management is critical to survival in the now extremely arid environment.

On a recent visit to Namibia, after travelling for hours in our four-wheel drive convoy across sandy plains and over barren, rocky landscapes shimmering in the midday heat, we arrived at a particularly colourful area – not from life, none was visible, but from the rocks coloured red, pink, purple and green due to their mineral deposits. I declared, 'This must be one of the most harsh and remote places on Earth!' Our guide and chief of the Topnaar community, Len Kootjie, then pointed out that his mother lives just over the outcrop. Of Khoikhoi descent, the Topnaar community has a government concession to operate and live in the area, which is a national park. We went to visit her and chatted in Afrikaans about her daily life there, about borehole flows, a rogue leopard interfering with livestock such as goats, and upcoming trips to town for supplies.

Prior to the arrival of the Khoikhoi people in the southerly parts of Africa around 2,000 years ago, the San people were widespread across the region, living as hunter-gatherers in a range of environments. Evidence of the ancestors of the San, potentially the first modern humans on Earth, dates back to 170,000 years ago along the southern coast of Africa. Traditionally, the San are a peaceful people, never having developed any weapons of war. Resource exchanges between communities took place without explicit agreements for compensation; the San lived in harmony with each other and their natural environment over countless generations.

In the past few thousand years, the San people were concentrated into more arid areas like the Namib Desert – the oldest San rock art in southern Namibia is radiocarbon-dated at 26,000 years old – by Khoikhoi pastoralists, Bantu farmers and then European colonists. Over centuries, colonisers committed genocide on the San people, killing or enslaving entire communities, and in doing so ending the San's migratory way of life across Southern Africa. Genetic studies of San people alive today corroborate that their population dates back about 200,000 years, while they show some of the highest levels of genetic diversity of any humans studied, suggesting that we all descend from their ancestors: the original modern humans. Today many San people, pushed into some of the harshest regions on our planet to eke out their existence, live in conditions of poverty as one of the lowest-income groups in Southern Africa. Yet within these ancient cultures lie the tools we may need to navigate the future. There are important lessons in resilience to be learned from a culture that has prevailed over most of the history of our species. When we look at other parts of the world with similarly long track records of inhabitants, we see similar patterns.

For example, between about 100,000 and 13,000 years ago the interior of the Australian land mass is thought to have been even more arid than at present, and it has been home to the Aboriginal people for over 60,000 years. Among other inherited abilities to withstand the desert's hot days, as well as cold nights, almost half of Western Australia's indigenous population have a genetic mutation in the regulation of the hormone thyroxine, reducing metabolic response to body temperatures of over 37 degrees Celsius (while saving energy for cold nights): advantageous in environments where temperatures regularly reach the 40s.

The arrival of the British in the 1700s spelled the end of their relative cultural isolation. Epidemics, seizure of land and water resources and genocide in the form of over 270 frontier massacres followed the arrival of the British, and while a government apology was issued in 2008, today Aboriginal people are more likely to suffer from poverty, unemployment and low life expectancy than other Australians, and they continue to fight for reconciliation and the right to live on their traditional lands.

While a lack of advanced technology, and indeed any weapons of war, meant that these communities were soft targets for marauding and typically genocidal colonists, there are lessons we can learn from what remains of their societies. Both of these ancient cultures managed to adapt to living, and indeed thriving, in extreme conditions. For tens of thousands of years the San and

Aboriginal peoples shared semi-nomadic lifestyles, living in egalitarian communities without formal leaders, with an emphasis on group consensus in decision-making, a respect for generations gone by and an economy based on sharing of resources, free of notions of ownership. They lived self-determined and self-sufficient lifestyles with little impact on their environment, valuing pacifism and equality, which is extended to other humans, animals, plants and their land. They revered, and saw themselves as one with, the natural world around them.

Such cultural characteristics enabled resilience in extreme environments over vast time periods. How far current global mainstream culture has deviated from these values is something we may consider in the face of extreme times ahead, but off-world communities will have the opportunity to rethink culture from the outset.

Off-world analogues

The people who travel to live and work in Earth orbit, on the Moon, Mars and beyond will have many tasks to take care of, while being exposed to the risks and dangers inherent in being beyond Earth in an environment far away from other communities and limited in liveable space. There exist a range of studies of people working in isolated, confined and extreme environments which provide analogues of aspects of off-world settlements. Polar stations and expeditions, deep-sea dives, hyperbaric chambers, submarines, simulated space capsules, underground bunkers and purpose-built analogue habitats are examples of environments which are comparable with conditions beyond Earth and which have contributed to studies of what off-world communities may experience.

HI-SEAS (Hawai'i Space Exploration Analog and Simulation) is a habitat in Hawai'i at approximately 2,500 metres above sea level. The HI-SEAS habitat sleeps a crew of six, is semi-portable and includes a kitchen, laboratory, bathroom, simulated airlock and work area. From the site the visual impression is one of isolation, yet accessible by a dirt road are a hospital and other emergency services within one hour's driving distance, or much less by helicopter.

The Mars-like landscape allows crews to perform geological fieldwork on weathered basaltic materials similar to Martian regolith. It is surrounded by relatively recent lava flows with very little plant or animal life present. The flows include a wide variety of volcanic features to explore, such as lava tubes. Furthermore, Mars-like twenty-minute delays are imposed on the communication systems between the crew and mission support, with no real-time conversations.

The Mars Society's Mars Analog Research Station (MARS) Project develops key knowledge to prepare for the human exploration of Mars: training field tactics based on environmental constraints like working in spacesuits; testing habitat design features and tools; and assessing crew selection protocols. The Mars Desert

Research Station is situated in southern Utah, 12 kilo-metres by road from the nearest town, in a region with a Mars-like terrain and appearance. Other analogue research stations built by the Mars Society include the Flashline Mars Arctic Research Station on Devon Island in Canada, the European Mars Analog Research Station in Iceland, and the Australia Mars Analog Research Station, currently in the planning stages.

These and other similar simulation projects provide important data on crews living and working in environments with some features in common with what we can expect off-world, as well as limited testing opportunities for the kinds of life-support technologies required beyond Earth. However, these kinds of simulations don't capture what is arguably the most mission-critical factor: the psychology. To learn from how people interact with each other and with their technology in an extreme environment requires a truly extreme environment. Building infrastructure and team spirit in the harshest conditions on Earth raises such experiments above simulation.

There is, however, one space-simulation mission that went further than any before or since: Biosphere 2. Originally constructed between 1987 and 1991 in Arizona, it is a 1.3-hectare structure built to be an artificial, materially closed ecological system to demonstrate the viability of human life in outer space. It remains the largest closed system ever created. Its seven biome areas consisted of a rainforest, an ocean with a coral reef, mangrove wetlands, savannah grassland, fog desert and two anthropogenic biomes – an agricultural system and a human habitat with living spaces, laboratories and workshops. Heating and cooling water circulated through independent piping systems below it, with solar heat and light entering through the glass frame panels covering most of the habitat; power, however, was supplied to Biosphere 2 from an onsite natural gas facility.

In the long term, it was seen as a study of closed biospheres towards human settlement of space; immediately, it allowed the study of ecology through manipulation of a small-scale biosphere. Two closed-system experients were performed in Biosphere 2: from 1991 to 1993, and from March to September 1994. These implementations ran into challenges including low amounts of food and oxygen, die-offs of animals and plants included in the experiment, tensions among the resident team, as well as external politics and media and disruption to the leadership of the project. However, the experiments set world records in closed ecological systems, agricultural production, health data with the low-calorie, high-nutrient diet the crew followed, as well as insights into the self-organisation of complex biomic systems and atmospheric dynamics. The second closure experiment achieved food sufficiency and did not require injection of oxygen. Biosphere 2's artificial sea became an extraordinary model for the effects of ocean acidification. Biospherian Abigail Alling observed that when carbon dioxide levels rise, ocean pH plummets and corals and other calcareous organisms perish – remember the Great Dying?

While I was in Bali to speak at an event, I spent the day with the late Biospherian and agriculturalist Sally Silverstone. She emphasised how much work was required to produce food; that each team member needed to be prepared to invest several hours a day in agricultural labour for the sustenance of the group. We visited fellow Biospherians marine biologist Abigail Alling and engineer Mark Van Thillo, who continue to live the dream of mending the divide between humans and the natural world on the Biosphere Foundation boat, which was docked in the harbour. I asked if they had homes on the mainland and they said they did not; they live permanently on the ship. In response to questions on the major take-aways from the Biosphere experiments, they said that team dynamics will always be the make-or-break of missions into the unknown; that having a team of eight, for example, resulted too easily in a split into two groups. While each person may be single-minded in performing their particular research programme, they emphasised that having a unity of purpose and an ability to put the well-being of the group ahead of one's own objectives is crucial to success.

The Biosphere Foundation was established in 1991 for research and education, with the mission of mitigating the effects of climate change through ecological restoration to ensure the right for all people to have clean air, water and food. The Foundation runs learning programmes, often with schoolchildren, on how to steward ecosystem health and biodiversity, reclaim water, recycle waste, restore soil health and grow food at home, with the aim of changing the way we perceive our relationship with the world and keeping 'biosphere love' alive in every head, heart and action. Biospherian and ecologist Mark Nelson has explained that our current ideas about the biosphere feed into an erroneous, dualistic, dyadic vision in which we humans are separate from the Earth, and that stepping back from the illusion of this separation is the beginning of understanding what it means to be a Biospherian. We would do well to keep the torch of the Biosphere philosophy alive as we enter an era when off-world expansion is within reach.

5

INNOVATORS

Leaving the cradle

Wherever on the surface of the Earth we happen to currently live, our ancestors' journeys into the unknown over the past 200,000 years made this existence possible. We have survived supervolcanic eruptions, asteroid impacts and ice ages, traversing mountains, deserts and oceans, equipped with our survival tools and each other. Throughout our history we have gazed at the horizon and set out to see what is there; on foot, on horseback, by seacraft, and later by train, car and aeroplane, we have migrated far and wide across the Earth.

Much of this happened before the existence of communication systems, and knowledge about these voyages away from home would have been passed on orally within and sometimes between communities. The impact of the journeys, however, whether experienced first-hand or through storytelling, without fail served to fire imaginations, broaden perspectives and advance our understanding of the world and our place in it.

With motivations ranging from survival to 'because

it's there', our exploration of the surface of our planet has left few stones unturned; from the highest mountains to beneath the ocean, from the driest deserts to the frozen poles, the human drive to experience all aspects of our world has characterised our time here. And as our technological capabilities increased, curious human eyes moved from the horizon to the sky, as the stage on which the next era of exploration would play out. Curiosity has been a major driver of space exploration, and the quest to understand the environment in which our planet exists has already dramatically advanced our perspective of our world; the view of Earth from orbit, for example, has become critical to our societal functioning.

Exploring beyond the boundaries of what we know results in new ways of perceiving our world. Today we stand on the brink of expanding our home beyond Earth, the culmination of 4 billion years of evolution! And if anyone is indeed watching the intergalactic *How to Grow Intelligence from Scratch* reality show, our evolution from single-celled organisms to beings on the brink of coordinating a multiplanetary existence certainly promises to be a grand finale.

A pivotal moment for our human psyche, the future of space exploration and modern communication systems came all at once in 1957, with a 'deep beep-beep' in the key of A-flat emanating from a transmitter aboard our first ever satellite, audible by shortwave radio and also occasionally visible to the naked eye. Sputnik was the first piece of human technology to operate in space, in orbit around our planet. After successfully launching Sputnik in 1957, achieving ninety-six-minute orbits with speeds of nearly 30,000 kilometres per hour, the Soviets wasted no time in building an additional pressurised compartment. Later the same year, the second-ever orbiting spacecraft carried the first dog, Laika, into orbit, although sadly not back as automated de-orbit systems were yet to be developed. Man's best friend had demonstrated that launch into space, as well as the weightlessness and increased radiation experienced there, is not deadly to mammals, paving the way for all future human space exploration. And at breakneck speed, the Soviets proceeded in taking the next steps.

With unimaginable courage, in 1961, just four years after the launch of our first satellite, Earth's first cosmonaut Yuri Gagarin declared: 'In all times and all eras humanity's greatest joy has been to take part in new discoveries. Let's go!' He then became the first human to journey beyond Earth into space; to experience the weightlessness of being beyond Earth's gravitation field; and to see the whole planet at a distance below. 'Orbiting Earth in the spaceship, I saw how beautiful our planet is. People, let us preserve and increase this beauty, not destroy it!' he proclaimed. Gagarin orbited Earth once at an altitude of around 300 kilometres in his five-ton spacecraft before landing safely back in a field in the Soviet Union by parachute almost two hours later. A farmer and her granddaughter observed the strange scene, and

Gagarin recalled: 'When they saw me in my spacesuit and the parachute dragging alongside as I walked, they started to back away in fear. I told them, don't be afraid, I am a Soviet citizen like you, who has descended from space and I must find a telephone to call Moscow!'

No less mind-blowing was an event that occurred later that same decade, with humans setting foot on the brightest object in the night sky, the Moon. For the first time we were seeing, up close, that familiar companion to our planet that we have gazed up at in wonderment from afar for millennia. I arrived for lunch at an air show in Bahrain a few years ago, and was over the moon to see Apollo astronaut Al Worden seated at the table to which I was directed. He maintained a special light in his eyes, and an excitement for humanity to get not only to Mars but beyond. He dedicated much of his later life to promoting a renewed space programme, education in the sciences and talking to children about his experience until his death at age eighty-eight in 2020.

Worden was the command module pilot for Apollo 15, the fourth successful crewed Moon landing mission, and spent three days alone in orbit around the Moon – physically, the most alone anyone has ever been (he still holds the Guinness World Record) – while crewmates David Scott and Jim Irwin explored the lunar surface. 'Now I know why I'm here. Not for a closer look at the Moon, but to look back at our home, the Earth,' he said.

The overview effect – a cognitive shift reported by astronauts while viewing the Earth from space described as 'a state of awe with self-transcendent qualities' – was not limited to those who left our planet. Data shows a spike per capita, particularly in the US, of students enrolling in science and engineering programmes as a result of being alive during the Apollo era and seeing the unfathomable being achieved through imagination, dedication and technology. A generation of inspired dreamers gave rise to the subsequent development of the personal computer, mobile phone and, my personal favourite, the Internet.

While we have not been back to the Moon since 1972 – although there are plans to return in the next couple of years – we have a legacy of over two decades of permanent habitation in Earth orbit, as well as having developed our technological capabilities at an unprecedented rate since then. It is now time once again to expand our horizons beyond the world as we know it; space exploration is a way to understand where we come from, who we are, and to dissolve the artificial boundaries between our nations and each other here on Earth. It is also a celebration of our humanity, an expression of our curiosity and drive to create and innovate. Centuries from now, there may well be human Martians telling tales of the perilous journey their ancestors made in the twenty-first century from Earth. Or, as a Tanzanian girl told me when I gave a talk at her school, 'I see it! Someday going to Mars will be like taking a flight from Dar to New York.' Historically speaking, there are reasons to believe that progress at this new frontier

– due to the novel conditions on Mars, and the kind of pioneers that life there will attract – will outpace developments happening within the system back home, and we may find ourselves in a new paradigm dominated by Martian technology and ideology. But we're jumping ahead. First, let's delve into what exactly it will take to get there, and whether we have the right stuff.

Contrary to popular belief, space is not far away. We are, in fact, in space, hurtling around our star at over 100,000 kilometres per hour. Even for those who don't travel often, 100 kilometres is not such a distance – that's where space starts – and just around 400 kilometres above us is where the Chinese Space Station and the ISS zip around our planet.

We are also currently in the midst of an era of rapid technological development, enabled by connectivity and data and characterised by new technological capabilities at the intersection of a range of fields. Space exploration has played a fundamental role in this development; many of the technologies we find so indispensable today were developed specifically for the challenge of exploring beyond Earth. Ground-based and remote exploration of space has led to massive progress in fields from computing to robotics. Imagery of our planet, people living in orbit around it, a global communications network and positioning system, and perhaps the very notion of globalisation itself, all rely on a perspective of our planet enabled by technologies operating beyond Earth.

Establishing a human presence on the Moon and Mars in the coming decades promises to further revolutionise our capabilities. For the first time in the 4-billion-year history of life on this planet, we have begun our expansion beyond Earth. What a time to be alive! We've come a long way since harnessing fire a million years ago. When considering the specifics of where we might explore next – and crucially, *how* – there are a few key developments that our aspirations in exploration have given rise to: exploration-driven innovation, as I like to call it.

Communication: starring the photon

Light is central to our understanding of the world. My mom tells me it was my first word; the beginning of a lifetime of peering into what light can tell us about the reality in which we find ourselves. Made up of massless and neutral particles called photons, and also manifesting as a wave with the properties of wavelength and frequency, it's the fastest thing we know of. Light can carry information and also travel relatively unimpeded through a variety of environments, and has therefore become the primary way in which we send and receive information around Earth, and also beyond.

Based on principles from the field of optics, where reflection, refraction and colour were described as early as the second century by mathematician Claudius Ptolemy, Galilei produced telescopes with over twenty times magnification in the early 1600s, enabling astronomical observations of Jupiter, surface features on the Moon, the phases of Venus, as well as spots on the Sun. At that time, however, the underlying nature of light itself, whereby all observations are made, had not yet been revealed.

During my second year at university we had been busy in the lab with strings and clamps and mechanical oscillators, activities that I found rather tedious – this and the requirement to wear closed shoes in the experimental physics laboratory playing a crucial role in determining my future as a theorist. Following the experiment, we had a lecture where we were reminded of the wave equation that we had derived in the lab; the mathematics describing waves like those we were generating on the string. Except that the lecturer wasn't talking about strings, he was talking about how a changing electric field generates a changing magnetic field, which in turn generates a changing electric field, and so on.

In 1864, physicist James Clerk Maxwell predicted that this interaction would create waves of oscillating electric and magnetic fields, so-called electromagnetic waves. But just what these electromagnetic waves were, was yet to be discovered. Our lecturer then demonstrated that when we rearrange the wave equation describing this electromagnetic wave to solve for its speed, what we get is quite a large number. Not just any old large number, though: in fact, rather astoundingly, the number Maxwell first calculated from the known constants in the equation, the same number which by then our lecturer had written up on the blackboard, was remarkably close to measured speeds of a phenomenon thought at the time to be completely unrelated to electric and magnetic fields. The number was 300,000,000 (metres per second), and the phenomenon: light. While it was known that light can behave as a wave, Maxwell saw, for the first time, that light is an *electromagnetic* wave!

Above the noise of students leaving the lecture (which had just gone over time), with the professor still writing up the implications of what he had just shown us, time stood still. Covered in goosebumps, I looked at the girl next to me, quantum biologist Betony Adams, who hadn't moved an inch. She turned, and we shared wide-eyed looks of wonder: the intimate inner workings of the Universe had just been revealed, and it would never look the same again. Our lifelong friendship and love of physics was firmly established.

By the turn of the twentieth century, Guglielmo Marconi, from humble beginnings in his attic with the help of his butler and building on Maxwell's discovery, sent the first electromagnetic radio waves across the Atlantic Ocean, and in doing so established the basis of modern communication systems. The key to sending messages with electromagnetic waves is that information can be encoded in the light. Marconi used a switching system to produce Morse code; in modern systems, properties like frequency or amplitude of the wave are typically used. Visible light can also be used as a medium of communication, but, with wavelengths similar to the size of atoms, can't pass through most objects – for example, walls – unless guided through such barriers, for example in a fibre-optic cable. Lower-frequency, longer-wavelength radio remains effective for long-distance communications whether on Earth or beyond, owing to radio's longer wavelengths, which can pass more easily through obstructions like buildings, particles in the atmosphere or even interstellar gas clouds.

Exploration involves investigating novel locations and, importantly, being able to communicate what is discovered back home. As a significant upgrade from oral or written messages, with our new-found ability to encode information in electromagnetic waves and transmit messages at the speed of light we are able to send vast data sets from one place to another, containing information on what is out there. While humans have so far only travelled as far as the Moon, the Voyager missions are now over 19 billion kilometres away in interstellar space, continuing to send back signals updating us on their journey. In the past few decades, numerous space missions have sent information back to Earth on the characteristics of a range of planets and moons in the Solar System.

If Earth is in direct sight mission data can be sent directly, otherwise the signal can be relayed via other spacecraft. For example, rovers on the farside of the Moon or on Mars utilise spacecraft in orbit there to get information back home. On Earth, a range of large dish receivers positioned in various locations around the planet receive and then transmit these signals to the various teams on standby to receive the data. These dishes can span dozens of metres in diameter, because as signals travel through space they become weaker, requiring a large receiver to collect enough radio energy, and extremely sensitive receiving equipment to discern the information content.

Space communications have to date been achieved essentially by this form of point-to-point signalling. For a total of a few dozen space probes, this has been sufficient. Recently, however, more countries, as well as private companies and organisations, are developing space exploration capabilities, and various projects are aiming to send crews to Earth orbit, the Moon, Mars and beyond in the coming decades. The Internet has revolutionised the way in which we communicate, operate and think, and the extension of this network into space is a critical next step as we prepare to support human activity beyond Earth.

The nearside of the Moon is close enough to be integrated with the Internet here, while local networks on the surface of, for example, Mars would be able to run locally on the same protocols on which the web functions on Earth. However, for activities on the farside of the Moon, or for a network enabling data-sharing between probes or communities on Mars or beyond with Earth, the primary limitations are unavoidable delays and disruption of communications due to the finite speed of light relative to the large distances in the Solar System, as well as planetary rotation.

Vint Cerf, one of the founders of the Internet, continues to create networks, but this time off-world. He and his team are working towards building storage capability into each communications node as a way of addressing these issues, and the testing of such delay-tolerant networks in space is already underway. Listening to Vint present a few years back in Lindau, Germany, I was struck by the implications of an Interplanetary Internet – which may be established within our lifetimes. The benefits of such off-world data transfer networks for space exploration are huge, including, given that we are smart about mitigating interference for sensitive scientific instruments, for our own Africa2Moon radio telescope, planned for the lunar farside; furthermore, excitingly, the end users of the Internet in space could soon be human. From Earth orbit to the Moon, and perhaps soon to Mars, connectivity will be critical for people living beyond our home planet. But before we launch beyond Earth, let's take a look at how information storage, together with computing capabilities, made the Internet possible, and the role that space exploration has played in driving these advances.

Information processing: featuring the electron

The photon is the star of transferring information from one place to another, from providing us with information on the state of the Universe or the other side of our planet to enabling us to read the words written here. On the other hand, a particularly useful particle for processing such information is the electron: a negatively charged particle with a small but measurable mass. Through control of the flow of electrons, or current, through a circuit, electronic devices can process data. But let's go back a few steps and look at the history of representing data in physical systems.

Beyond inherited DNA, where information for bodily development and function is found, the information that we learn about the world during our lives was originally stored in human memory; the development of language revolutionised our ability to categorise and pass this information on orally within and between communities. Later, with written language, we could store information outside ourselves, for example in manuscripts or maps. Information processing, for example counting, also began to move into the external world, and from using our fingers we developed counting devices. The Lebombo bone found in Southern Africa may be the oldest known mathematical artefact, dating from around 40,000 years ago and consisting of twenty-nine distinct notches etched into a baboon bone; the abacus is a more sophisticated counting device dating back thousands of years (and still used by at least one elderly lady in a tiny village in Japan to tally my purchases from her shop).

Fast-forward to the early 1800s, when the punch card was invented to programme looms for weaving textiles, with each row of punched holes corresponding to a row of a textile pattern. Extending the concept that a unit of information can be encoded in the presence or absence of a hole on a card, which allows current to flow (or not) between two conductors separated by the moving card, computing pioneer Charles Babbage first described in 1837, but never built, the first known digital programmable computer, whose structure has prevailed in modern electronic computer design: the Analytical Engine. Mathematician Ada Lovelace saw the potential for such a machine, able to store both data and a program, or sequence of operations, which she already understood to go far beyond calculation; she described an operation as representing any process altering the mutual relation of any two or more things in the Universe. Her notes of the early 1840s included an algorithm designed to be carried out by the machine – considered to be the first computer program – and envisaged that the engine might compose music or images: 'We may say most aptly that the Analytical Engine weaves algebraical patterns just as the [programmable] loom weaves flowers and leaves,' she said.

Then, in 1884, statistician Herman Hollerith invented, and built, an electromechanical tabulating machine to capture and process information stored on punched cards, thus establishing the foundations of digitisation and data processing. Digitisation, meaning representation as a string of zeros and ones, makes storing, processing and transmitting data in physical systems both universal and efficient. Just as the encoding of messages in the properties of light waves revolutionised communication, in digital electronics representing zeros and ones by the presence or absence of an electrical signal gave rise to modern information processing and computing.

Early computers used vacuum tubes to control electric current for circuitry, and magnetic drums where pulses of current could be stored for memory, which were huge by modern standards and used a lot of power. For example, the set of computers called Colossus, built to decrypt encoded messages during the Second World War, was the first programmable, electronic, digital machine. It was manually programmed by switches and plugs, and used thousands of vacuum tubes and a series of pulleys transporting rolls of punched paper tape to reduce decryption time from weeks to hours; it also weighed a few tonnes, took up an entire room and required hundreds of people and kilowatts of power for its operation.

In 1947, the first known transistor was demonstrated by John Bardeen, William Shockley and Walter Brattain, who were later awarded the Nobel Prize in Physics for their discovery. A transistor is an electronic component containing a semiconductor material that can either conduct or insulate electric current, meaning it can be used as a switch in electrical circuits. A transistor is the fundamental building block of computer circuitry: the transistor either prevents, '0', or allows, '1', current to flow through. During the 1950s, semiconductor devices gradually replaced vacuum tubes in digital computers, and in 1953, what is thought to be the first operational transistor computer was demonstrated by Richard Grimsdale and Douglas Webb.

Transistors conducted electricity faster and better than vacuum tubes, and transistor machines were orders of magnitude smaller in size and power requirements. And in 1958, less than a year after the launch of the first satellite, Sputnik, in 1957, Jack Kilby demonstrated the first combination of multiple interconnected electronic components elements like transistors, resistors and capacitors in a single integrated circuit, or microchip. As the fundamental building block of all modern electronic devices, the potential for device miniaturisation, enhanced functionality and reduced cost ushered in the era of modern computing. The next step was to connect these computers.

In what may have been the first proposal for a computer network, in a 1960 paper psychologist and computer scientist J. C. R. Licklider wrote: '[With] a network of such centers, connected to one another by wide-band communication lines ... the functions of present-day libraries together with anticipated advances in information storage and retrieval and symbiotic functions [can be achieved].' He had no lack of vision either, writing memos to his team addressed to 'Members and Affiliates of the Intergalactic Computer Network'. Bringing Licklider's concept to fruition, the first message was sent between two computers a few months after the first Moon landing in 1969 as part of the Advanced Research Projects Agency Network, or ARPANET. Monitoring the computer transfer while also on a phoneline between the two locations, computer scientist and head of the ARPANET team, Leonard Kleinrock, said in an interview: 'We typed the L and we asked on the phone, "Do you see the L?" "Yes, we see the L," came the response. We typed the O, and we asked, "Do you see the O?" "Yes, we see the O." Then we typed the G, and the system crashed. Yet a revolution had begun.'

ARPANET grew, and by the early 1970s included dozens of nodes. But getting computers to communicate – networking – was only the first step; now we needed to get networks to talk to one another: so called internetworking. And in 1973, the so-called fathers of the Internet, Vint Cerf and Robert Kahn, began the design of the Transmission Control Protocol (TCP) and the Internet Protocol (IP) which standardised data flow and laid the foundations of the TCP/IP protocols that enable the Internet today.

By 1971, the first digital network email was sent on the ARPANET by computer engineer Ray Tomlinson to a computer next to his; it said 'something like "QWERTYUIOP"', heralding the arrival of the global communications network that plays a central role in information-sharing today. Meanwhile, researchers at what was and remains one of the largest data generators on the planet, the particle collider at CERN (soon to be exceeded by the SKA radio telescope when operational in the next few years), were faced with challenges that included storing, updating and finding documents and data files associated with the project. To solve these organisational issues, scientist Tim Berners-Lee invented the World Wide Web (WWW) in 1989, elevating the capabilities of the early computer networks beyond the sending of messages and into the realm of scalable data-sharing.

And scale it did! Today, almost 20 billion devices are online. This means anyone with an Internet connection can achieve amazing feats like reading any book, accessing supercomputing resources remotely, or zooming down to a recent street view of almost any city on the planet. To date, we have created, communicated and consumed almost 200 trillion gigabytes of data. This is predicted to double in the next couple of years. Almost half of the data we currently produce comes from machines: devices that are able to sense the environment, from microphones to satellites in Earth orbit. With the help of our personal devices, we output the rest. A few years ago, it was estimated that 90 percent of the world's data was generated in the past two years; and the ratio between annual data production and total human data continues to grow. The benefit of all this information is determined by our, or increasingly our machines', ability to analyse it.

Typically, (raw) data needs to be organised in order for it to become useful to us as information. The algorithms that enable us to learn from data have become more and more complex; and a rapidly growing field, spanning many areas, is that of data analysis. Data processing is the task of converting raw data into more meaningful information that can then become knowledge. Current state-of-the-art machine learning algorithms, often referred to as artificial intelligence, can be automated such that we are typically not aware of the actual algorithms being employed. Nonetheless, through repetition and large data sets, underlying patterns embedded in the data can be uncovered. Our data processing and computing capabilities have impacted almost every aspect of our lives; let's look at space exploration in particular.

By the time we were ready to go to the Moon, computing still left a lot to be desired by modern standards: physicist Katherine Johnson did calculations by hand for the first Moon landing in 1969. However, crews arriving there were equipped with computers that flew most of Project Apollo, except briefly during landing. Only in the 1970s did the first generation of home computers reach comparable performance levels to the Apollo Guidance Computer, which, although cutting-edge at the time, had less computing power than a modern USB-C charger. Putting computing into machines was an important next step for exploration.

Robotics is a field that brings our ability to program computers into the physical world; we can build and program a robot to perform tasks that eliminate the requirement for human labour, and also tasks that are dangerous or impossible for humans. The word 'robot', from the Czech word *robota*, meaning forced labour, was first used in 1921 by playwright Karel Čapek in a play about mechanical factory workers that rebel against their human masters; then, in 1961, inventor George Devol contributed to founding the robotics industry by developing the first digitally operated programmable robotic arm, installed at General Motors to retrieve and stack hot metal pieces for car production. Robots can manufacture things, even robots. Critically for space exploration, they can survive the extreme conditions beyond Earth with greater predictability and resource efficiency than humans could.

Since HAL 9000, a fictional artificial intelligence character created by Arthur C. Clarke and featured in Stanley Kubrick's movie *2001: A Space Odyssey* (1968), the idea of robotic assistance in space has prevailed in science fiction, and more recently in reality, as the rovers on Mars assume personalities on social media, taking selfies and capturing the public imagination by sharing visual stories from 200 million kilometres away.

For humans, space is an extreme place; and while we have only been as far as the Moon, our direct knowledge of a range of locations in the Solar System is thanks to robotic missions that have travelled there. Sputnik was the first robot in space, while the Voyager missions achieved the milestone of taking robotics out of the Solar System into interstellar space. Robots don't require the complex resources that humans do, and as long as their power, controls, sensors and communication systems are functional, they continue working in all kinds of harsh environments reliably and without rest. Largely because of the data acquired on the surface of Mars by more than two decades of roving with a combination of automated and remotely operated robotic vehicles, can we begin planning the first crewed missions there.

Rovers on Mars are extreme examples of so-called edge computing, whereby data is processed locally in a remote location, closer to the source of the data, and far from the centre of the network. Currently active on Mars are the Curiosity and Perseverance rovers, with China's first ever Martian rover, Zhurong, having ceased to function in May 2022. Signals from Mars take on average more than ten minutes to travel back to Earth, so the rovers are equipped to make some decisions themselves, enabled by their ability to compute. Both NASA rovers have processors similar to those used in desktop devices from the late 1990s with only 10 million transistors, which is about 1,000 times fewer than your average smartphone chip. The reason for this is that the computer needs to be robust against the higher levels of radiation on Mars, and perform reliable error correction to repair any damage to data in the memory. Perseverance has three computers on board, with two gigabytes of flash memory and 256 megabytes of RAM each, two of which take care of the rover's main functions and analyse navigation images, with the third acting as a backup.

The combination of robotics – from simple sensors to humanoids – with predictive data processing techniques is particularly useful in the context of extreme environments on Earth or beyond: from air or spacecraft functionality, to resource availability or life-support system health, to physiological indicators. Monitoring a plethora of parameters to build large sets of data, together with machine learning techniques, allows the prediction of errors before they occur, vastly increasing the safety of people and the performance of the equipment keeping them alive.

Where to next? The number of transistors in processors has been increasing exponentially since the invention of computing; modern processors contain billions of transistors to perform increasingly complex digital calculations. Moore's Law tracks the associated increase in computing power as a function of time: the speed and capability of computers has been doubling every two years. However, we are reaching a fundamental limit to the number of transistors we can pack in; not because of technological limitations, but rather because the laws of physics will prevent transistors from functioning traditionally if they are any smaller and closer together. We are approaching the classical limit of computing; can quantum mechanics provide the tools to do computing on even smaller scales?

A few years back, I had a bizarre dream: I found myself in a world of few colours, with a black background, organised in a three-dimensional grid. In this dream, I am on one of the grid points. I have limited degrees of freedom, a vector with three-dimensional rotation about a fixed point at my feet. A ripple of information in the form of a band of colours is incoming from a distant region. Each vector responds and swings into alignment as the glow of the signal approaches. I get ready to respond.

The day before I had that dream, I had been enjoying the Singapore nightlife beside one of the city's large canals with some friends and fellow researchers. I was a PhD student abroad, fairly broke and full of bravado, and it was very warm, even at 2 a.m., so when someone dared me to swim across the roughly 40-metre canal and back for 200 Singapore dollars I accepted. I slid down the slimy stairs into the totally opaque brown water.

I did my best not to swallow any water during the fairly long swim, rather disturbed by the slow but persistent current and thinking of potential objects beneath me, as well as the possibility of being arrested! Nonetheless, the next day, and well into the night, I had one of the most intense fevers of my life. Sweating, tossing and turning, I envisioned that I was a qubit in a quantum computer.

The project I had come to Singapore to work on involved quantum state transfer, and in my dream I became one of the many network nodes, part of a transfer event constituting some kind of computation; not only had I seen how quantum information processing works,

I had participated in it! I used the money to explore Borneo the following weekend. More importantly, we later applied our results in quantum state transfer to the photosynthetic complex of green sulphur bacteria, with implications that a noisy environment could assist rather than hinder aspects of quantum computation.

Quantum computing offers potential solutions to the limitations that traditional computers face. It can significantly speed up the timeframe for solving certain types of problems, and without running into the physical limitations that will eventually prevent the exponential growth we have been enjoying so far in our transistor-based computing capabilities; furthermore, it gives us the ability to solve certain problems that haven't been solved before. By harnessing the unique properties of the very small, quantum computers are able to transmit quantum states between qubits in a circuit. As in my dream, a qubit can be represented as a vector fixed at its base that can not only point up or down, which would correspond to '1' or '0', but also rotate in three dimensions to point in any direction around the fixed base, which corresponds to various combinations of a '1' and '0'.

A qubit is the basic unit of information in quantum computing in the same way that a bit is smallest unit of data that a traditional computer can process and store. Here, the qubit being in a probabilistic combination of *both* '0' and '1', rather than *either* '0' or '1', also called quantum superposition, allows quantum processors to perform multiple calculations at the same time on the same input signal, in some cases enabling exponentially higher-speed calculations than what is possible in traditional computing. While transistors can represent bits, in quantum computing a quantum object like an electron can be used to represent a qubit, with its spin being some combination of 'up' and 'down'. Photons can also be used to represent qubits in so-called optical quantum computing, which has advantages like room temperature functionality and integrability with existing fibre-optic-based telecommunications infrastructure.

The major application of quantum computing would be in modelling complex quantum systems. We've had a look at how challenging it is to model a single bacterial cell with a traditional supercomputer; understanding how sequences of amino acids result in functional three-dimensional protein structures, or protein folding, is another complex biological process on which quantum computing could shed light. In fact, some headway in solving the protein folding problem has recently been made using a combination of classical and quantum computing techniques. Further potential applications of quantum computing include the modelling of complex phenomena like global climate patterns or complex molecule interaction towards, for example, materials or pharmaceutical drug design, as well as for data processing techniques like machine learning, which have a plethora of applications in pattern identification from image recognition to risk assessment.

Talking about risks, in this description of the emergence of computing and connectivity we cannot leave out a third critical category of development, namely cryptography. The benefits of global (and interplanetary) connectivity come with associated risks around how the information that we create, communicate and store can be intercepted, sometimes with malicious intent. For people living in extreme and otherwise deadly environments, information concerning life-support systems should be tamper-proof. Cryptography is the ancient art of achieving confidentiality by transforming a message so that it is only intelligible to someone in possession of a key. Since the emergence of the Internet a multitude of algorithms for data security have been developed, and global standards for encryption protocols provide some level of communications security over computer networks. Satellite data transfers are typically encrypted, and communications with the ISS are protected by a range of protocols. Data relating to life-support systems of off-world communities would certainly need to be safeguarded.

We are not there yet, but a universal quantum computer, if developed, will be able to crack traditional encryption methods. For example, while a classical computer would take around 300 trillion years to break an RSA-2048-bit encryption key, a quantum computer could do this in a few seconds: a beautiful example of the exponential speed-up quantum computing provides for certain problems, but also an indication that if we value information privacy, for securing off-world life-support systems and data processing, for example, then we will need to prepare.

Fortunately, quantum encryption technology is a rapidly developing industry. Quantum physicist Artur Ekert, also one of my PhD mentors and my first scuba-diving instructor, proposed that the uncanny correlations shared between quantum objects under certain conditions, so-called quantum entanglement, can be used as a resource to generate a secret key. Thanks to quantum physicist and founder of ID Quantique Gregoire Ribordy and his team, and other more recent start-ups around the world, the technology has come a long way and is now commercially available. In my Samsung smartphone from Gregoire, ID Quantique's quantum random generator contributes towards enhanced security of the device, and a number of organisations around the world are already using quantum key distribution technology to secure data transfers against the projected capabilities of quantum computing.

It has already been shown that a hypothetical universal quantum computer – a general-purpose quantum computer that can implement any quantum algorithm, which remains beyond reach for current state-of-the-art quantum technologies – would easily be able to crack the traditional encryption systems still widely used to keep data private. However, in the drive towards this ultimate machine, only recently, and not without controversy, have announcements of achieving so-called quantum supremacy been publicised. Most recent claims by Google to have completed a calculation almost instantly, which would otherwise have taken up to fifty years, have again been met with the criticism that the best real-world application of this calculation is to confer bragging rights. While our best quantum computing technologies may be close to outperforming our best (traditional) supercomputers at real-world problems, many challenges, particularly in error correction, lie ahead.

Let's return to the relation between quantum effects in photosynthesis and quantum computing. Understanding how living systems such as photosynthetic organisms – which are complex, relatively warm, constantly evolving and interacting with the environment – can sustain rather than destroy fragile quantum correlations has important implications for artificial photosynthesis and more efficient solar cells, but also for the engineering of quantum systems to perform tasks such as quantum computation. The effects of the environment inevitably induce errors in the delicate quantum systems in which information is stored – one of the most serious challenges in the drive towards realising the quantum computer. Theoretical models inspired by photosynthetic light-harvesting systems have been proposed, showing how interaction with a busy environment with just the right features can enhance rather than hinder certain delicate quantum processes. However, our engineering remains rather rudimentary when we compare it with the intricacies of living things.

Quantum computing has recently become a buzz phrase, particularly in the corporate world; however, in its current form it will not replace traditional computing, but rather potentially augment it for a particular set of tough problems. The timeframes for achieving this general capability of fault-tolerant quantum computing are unknown, while some researchers in the field continue to question whether it's even physically possible. We can dream about the role quantum computing could play off-world, and consider lessons from our investigations into the nature of life on small scales that could assist with solving error correction challenges; however, we don't need a quantum computer to expand beyond Earth. But we do need a launch vehicle.

Getting off the ground: propulsion systems

Transportation systems are fundamental to space exploration. The first known rockets were used in China to propel so-called fire arrows, projectiles consisting of a bag of gunpowder attached to the shaft of an arrow, possibly more than 1,000 years ago. According to one written account, aviator Lagâri Çelebi made a successful manned rocket flight in 1633 in a seven-winged craft using 64 kilograms of gunpowder in Istanbul. Inspired by the novels of Jules Verne, in the late 1800s rocket scientist Konstantin Tsiolkovsky established the field of astronautics and pioneered a range of then-fantastic ideas, many of which have since been realised, while some may yet come to be – including multistage boosters, space stations, airlocks, life-support systems, hovercraft and space elevators.

In 1903, Tsiolkovsky proposed a method of reaching the minimum speeds – almost 30,000 kilometres per hour – required to orbit the Earth: with a multistage rocket that burns liquid hydrogen in liquid oxygen, creating the most efficient thrust of any known propellant. And just a couple of decades after his death in 1935, the Soviets put his ideas into practice and achieved the unthinkable by launching the first artificial satellite into Earth orbit in 1957.

Fuelled by the uncertainty of Cold War tensions, rapid developments in space exploration took place over the subsequent decades, including crews landing on the Moon and living in space stations in Earth orbit, as well as landers revealing surface conditions on neighbouring planets Venus and Mars. However, in terms of the spacecraft and fuelling systems used to get off the surface of Earth and back, we have not really reinvented the wheel.

A rocket is currently the only technology we have to get to Earth orbit or beyond (while Japan has announced plans towards implementation, Tsiolkovsky's space elevator has yet to take off). Traditionally, chemical rockets are used for this purpose: either solid-fuel rockets that originated with Chinese fire arrows and provide powerful thrust with a relatively simple design, or liquid-fuel rockets in the spirit of Tsiolkovsky that use liquid propellants such as hydrogen or kerosene together with oxygen and require smaller volume tanks. To get back to Earth, either a winged vehicle like the US Space Shuttle, retired in 2011, or a capsule like the original Soviet Soyuz – the longest-serving crewed spacecraft design still in use, by the US since 2011 and also in more recent craft designs by both SpaceX and China – is typically employed.

With existing technology, it takes around three days to get to the Moon, seven months to Mars and about a year to the Asteroid Belt. A speed of over 40,000 kilometres per hour is required to escape Earth's gravity altogether, at which point, limited by the mass of the fuel that needs to be lifted from Earth, propulsion systems are typically disabled and the spacecraft continues at constant velocity through the frictionless vacuum of space, saving any remaining fuel for adjusting course to get to the destination, and potentially to make a return trip. The mass of the fuel required to generate acceleration is the main limiting factor to fuel-based (including chemical) propulsion systems. How efficient can a fuel be?

Plasma rockets like Ad Astra's Variable Specific Impulse Magnetoplasma Rocket (VASIMR) are aiming for fuel efficiencies ten times those of the traditional chemicals employed. The VASIMR system uses radio waves to ionise a propellant into a plasma: a fourth state of matter where the electrons can wander freely between the nuclei of the atoms rather than being bound in individual atoms (recall the quark and electron plasma that filled the early Universe). A magnetic field then accelerates this plasma out of the engine, generating thrust.

Theoretically, VASIMR would be able to continuously accelerate through space and arrive at Mars in just thirty-nine days. Accelerating at Earth's gravity for the first half of the journey, reorienting the spacecraft, and decelerating at the same rate for the remainder would have the added benefit of enabling crews to experience Earth gravity for long-duration spaceflights throughout the Solar System. Former astronaut Franklin Chang-Díaz is leading the team endeavouring to bring VASIMR from the lab to flight.

A gram of nuclear fuel contains millions of times more energy than, for example, oil, but even accessing the heat generated by nuclear reactions could power spaceflight; nuclear thermal propulsion systems can as much as double the efficiency of chemical propulsion. To mitigate the risk of a nuclear explosion on launch, these kinds of propulsion systems could be launched into space by chemical rockets before they are turned on. Thermal nuclear propulsion is not a new concept; while a number of national and private programmes are working towards this capability, it hasn't yet been demonstrated in spaceflight.

Beyond chemical propulsion, there are some novel systems that can achieve better fuel efficiency; while useful in space, they are typically not powerful enough for launch. Recall the plans to accelerate a tiny craft to the nearest exoplanets in Proxima using a solar sail? While photons have no mass they do have momentum, and a spacecraft with large reflective sails could capture this momentum of light from the Sun for propulsion. Progress towards building sails large enough to propel crafts of significant size is underway. Ion propulsion systems are another highly efficient technology, using electrical energy from solar cells to provide thrust suitable for keeping communication satellites in position and propelling deep space probes, such as the Dawn probe to Ceres, where ion propulsion enabled entry and departure from orbit of the dwarf planet. An ion engine cannot generate sufficient thrust to achieve initial lift-off from any celestial body with significant surface gravity, though, and manoeuvres in space have yet to be tested for larger craft.

For the most part, crewed travel beyond Earth has, understandably, relied on tried and tested designs. In Kazakhstan there is a deep sense of national pride in the long-standing tradition of launching humans into space. Flying over vast, bleak uninhabited tundra, Air Astana's inflight entertainment screened documentaries on the country's history in space exploration and I began to understand the cultural importance of this activity, even before hundreds of children greeted me at their school in astronaut suits singing Kazakh songs about space. Until SpaceX first launched crew to the ISS in 2020 in their Dragon Capsule, all astronauts, cosmonauts and private citizens travelled to the ISS in Russian Soyuz capsules launched from Baikonur, Kazakhstan. The Soyuz was the primary form of transportation for humans to get to space after NASA's Space Shuttle programme was retired in 2011. Traditional propulsion technology works, and the Soyuz has been in use for half a century with very little modification. The Chinese launched the first taikonaut to space in 2003; the Shenzhou spacecraft employed for their crewed space programme is Soyuz-inspired, as is SpaceX's Crew Dragon. And in the current era of space exploration, we can't talk about rockets without talking about SpaceX ...

Entrepreneur Elon Musk had recently sold PayPal when he established SpaceX in 2002, with the goal of making spaceflight routine and affordable in order to make humans multiplanetary. I clearly remember watching the livestream in 2015, on the day when, while launching a cargo mission for NASA to the ISS, the booster stage of SpaceX's Falcon 9 rocket turned round, came back towards Earth and landed on a floating platform off the coast of Florida – achieving what traditional leaders in aerospace design had said to be impossible and opening up a new era in space exploration. By pioneering reusable rockets, SpaceX has already cut launch costs by an order of magnitude and become the first private company to deliver crew to and from the ISS.

For the real heavy lifting, SpaceX's fully reusable two-stage Starship is currently being developed. Capable of delivering over 100 tonnes of crew or cargo to the Moon, Mars and beyond and back, it will be the most powerful launch vehicle ever developed. Starship consists of a first-stage booster to leave Earth's surface, and a second stage to bring the spacecraft to Earth orbital velocity, which will also function as a long-duration spacecraft and lander. With over half a dozen test flights under the belt, Musk has announced that SpaceX will begin launching uncrewed Starships to Mars in 2026. But first, according to the current schedule anyway, Starship will travel to the Moon.

In 2021, SpaceX was contracted to build the lander for NASA's Artemis crewed missions back to the Moon. The launch vehicle, spacecraft crew module and the SpaceX lander form the primary spaceflight infrastructure for Artemis, with the Lunar Gateway space station in lunar orbit playing a supporting role. Originally planned for 2016, the uncrewed test of the Space Launch System (SLS) launch vehicle, together with the Orion space-craft, finally took place in 2022. Further lengthy delays included court cases with multi-billionaire Jeff Bezos disputing his company Blue Origin's exclusion from the lander contract award, and also various mechanical issues.

It remains to be seen whether Artemis can remain on schedule; in particular given the embattled Boeing's involvement in building the launch vehicle. Due to reg-ulatory and systemic organisational failures and the prioritisation of profit over safety, aircraft failures have caused the deaths of all people on board not one but two full Boeing planes, in Indonesia in 2018 and Ethi-opia in 2019; and more recently, critical failures have been reported of Boeing aircraft in locations around the world, both in flight and on the ground. Then only recently, in June 2024 (and not without significant issues like billions of US dollars' worth of overruns, as well as gas leaks and thruster failures that resulted in the crew having to manually operate the craft en route), did Boeing manage to launch people to the ISS subsequent to the signing of a multi-billion-dollar contract with NASA in 2014. Moreover, the two astronauts, Butch Wilmore and Sunita Williams, remained on board the ISS far beyond the planned eight-day mission; and are currently scheduled to return to Earth almost a year later with SpaceX in early 2025. For now, returning humans to the Moon – including 'the first woman and the first person of color', as marketed by NASA – to a location near the south pole has been pushed back to 2026. Exact dates for crewed launches to the Chinese International Lunar Research Station planned for later this decade remain to be announced. If at least one of these programmes is successful, in the next few years it will be all systems go for crewed exploration of the Solar System like never before.

Getting off the surface of Earth is just the beginning, though; living off-world requires reliable life-support systems. And decades of practice in Earth orbit, as well as a few brief excursions to our Moon, stand us in better stead than we may imagine.

The view from Earth orbit

Our planet is at just the right distance from the Sun to support liquid water, and life is a tenacious phenomenon here on Earth; the breathable air and nutritious food produced by networks of various organisms were available in sufficient quantities to support human communities in most places we explored. But in more extreme environments, those that do not support normal human function, more advanced tools are required.

In case the idea of expansion beyond Earth makes anyone nervous, this is an update to inform you that we have already expanded beyond the surface of our planet, a vantage point from which we have a convenient view of our world. While not very far from the surface of Earth, the continuous occupation of the ISS for the past twenty years, multiple Soviet stations of previous decades and more recently the Chinese Space Station are examples of the advancement of our society from beyond the surface of our home planet. The capability to support life in otherwise deadly environments has a history.

In 1405, military engineer Konrad Kyeser described a diving dress made of a leather jacket and metal helmet with two glass windows, with a leather pipe connected to a bag of air. This is the first record we have of an invention that would allow a human to survive – for a time, at any rate – underwater. In 1771, diving engineer Sieur Fréminet designed, built and tested a diving dress with a compressed air reservoir mounted on his back, using his invention successfully for more than ten years in harbours in France. In 1793, aviation pioneer brothers Joseph-Michel and Jacques-Étienne Montgolfier sent a duck, a rooster and a sheep up in a hot air balloon – the first of many aerospace experiments to use animals – with the intention of learning whether ground-dwellers can survive high altitudes.

In the 1800s, centuries of designs for underwater exploration culminated in the first crewed submarines, taking humans to as yet unexplored depths. Around the same time, the first steam trains were developed, and this too revolutionised what people believed about the conditions under which a human body could survive. People worried that they wouldn't be able to breathe or withstand the vibrations of such rapid movement, but within the next fifty years passengers were travelling at previously unbelievable speeds of 80 kilometres per hour. By 1938, passengers in commercial aircraft were breathing pressurised air 6,000 metres up at speeds of almost 400 kilometres per hour.

In 1961, when Yuri Gagarin became the first human in space, carbon dioxide scrubbing technology originally developed for use in submarines was demonstrated in space, and Gagarin was able to orbit Earth in the Vostok capsule for more than an hour. To safely eject from the capsule at a height of 7,000 metres with pressures around 40 percent of sea level, he wore a pressurised suit with closed helmet and gloves. Medical telemetry and communication would have been disconnected on ejection. He was able to breathe the air inside the suit for the short time of fast descent down to about 2,500 metres by parachute, without oxygen supply or carbon dioxide removal. At this height the air pressure is about 75 percent of normal, similar to the pressurised cabins of passenger airplanes, and Gagarin could open his helmet to breathe fresh air. A series of manoeuvres that Laika, unfortunately, would not have been able to make.

Today, the space above Earth's surface between altitudes of 160 and 36,000 kilometres is a shared resource for communication and observation, as well as research and development in the microgravity environment. Since the launch of Sputnik in 1957, thousands of artificial bodies have been sent into orbit around Earth: both operational and defunct satellites owned by a few private companies and the governments of almost 100 different nations on Earth, as well as space stations for human habitation launched and operated by a handful of countries.

The vantage point from above the surface of the Earth provides convenient monitoring and signal transmission capabilities, and satellites orbiting there are typically equipped with sensors to collect data on the atmospheric or surface conditions of our planet, or are relaying signals where point-to-point propagation of electromagnetic waves is obstructed. The benefits on the ground of satellites in orbit are huge: from providing reliable connectivity, in particular for communities disconnected from the world due to conflict, natural disasters or a lack of infrastructure, navigation systems, to monitoring systems collecting big sets of data on climate and weather patterns, the Sun's activities or natural resource management.

A range of space stations in Earth orbit have been an excellent opportunity to test life-support systems at a relatively safe distance from Earth; just a few hundred kilometres above the surface. Our first space station was launched into Earth orbit by the Soviet Union in April 1971, marking the tenth anniversary of the first human spaceflight. The Soviet Union launched eight space stations successfully, with the last, Mir, being deorbited in 2001. China launched two space stations as tests for key technologies, ahead of the launch of the currently operational Tiangong space station in 2021. The ISS has continuously housed an off-world community of six people at a height of around 400 kilometres above the surface of the Earth since 2000. People in space stations live in microgravity and are typically performing experiments or maintenance operations, with the support of large teams of people back on Earth. Protocol-based, detail-driven activities are planned meticulously; supplies, besides solar energy, are delivered from Earth; and evacuation from the ISS back to the Earth's surface is possible within a few hours.

Earth orbit has been home to many innovations as well as to humans. In 1982 the first plant was grown; in 1987 oxygen generation from water was first demonstrated in orbit; and in 1989 the first water recovery system was used to recycle water, primarily as a source of oxygen, both in Soviet space stations. Since its launch in 1998, over two decades of continuous human habitation of the ISS has been an opportunity to further sophisticate life-support systems in space, resulting in impressive research and development in areas of solar technology, water filtration systems, LED lighting for agriculture, additive-manufacturing, remote healthcare in extreme environments, as well as communication and computing systems, to name but a few.

A collaborative facility shared by space agencies from the US, Russia, Europe, Japan and Canada since 2001, and until the launch of Tiangong in 2021, the ISS has been the soccer-field-sized platform facilitating our entire planet's crewed off-world activities. Due to ageing after twenty years in service, coupled with government budget cuts, it was recently announced that the ISS will be decommissioned by 2031: deorbited and plunged into the Pacific Ocean as trash. There is discussion of transitioning habitation of Earth orbit to commercial operations, but plans to launch further space stations remain to be confirmed.

On the other hand, the over-500-billion-US-dollar global satellite economy – including telecommunications and Internet infrastructure, global positioning services, Earth observation capabilities, national security satellites and more – continues to experience massive growth. Decreasing launch costs have enabled a variety of industries to benefit from satellite technologies to drive innovation and efficiency in their products and services, as well as make a lot of money. For example, SpaceX plans to launch in excess of 40,000 satellites to complete its Starlink constellation, which at full capacity could provide the company with tens of billions of US dollars annually from millions of Internet subscribers all over the planet, thereby contributing to funding its Mars missions.

And as of 2024, there are more than 10,000 active satellites in orbit. If SpaceX, Amazon, Boeing, China and others achieve their ambitious constellation plans, there could be tens of thousands of additional satellites in Earth orbit within the next decade.

As usual, though, these advances come at a price. The accumulation of space debris, or artificial objects in Earth orbit that no longer serve a function, poses a risk to our activities in space. Space debris includes defunct satellites and spacecraft, abandoned launch vehicles, mission-related debris and fragments thereof. Of the Starlink satellites launched by SpaceX, to date hundreds are no longer functional – a further contribution to the estimated more than 40,000 pieces of debris larger than 10 centimetres currently orbiting the Earth.

Collisions between debris and operational satellites can disrupt observation and communication capabilities, threaten the safety of people in orbit, or in the long term trigger chain collisions that could spell the end of our ability to explore space altogether. The Kessler effect, proposed by scientist Donald Kessler in 1978, is a scenario in which the density of space debris in Earth orbit is high enough that collisions between objects cause an ongoing cascade of further collisions; the potential outcome being that launch and thus space exploration becomes near-impossible for generations. Our current economic system creates incentives for companies to innovate in areas where there are profits to be made. However, passively hoping that cleaning up Earth orbit, or indeed that the continued presence of people there will eventually be profitable and pursued commercially, seems counterproductive to say the least.

While India has committed to 'debris-free' space missions by 2030, additional regulation by international organisations may be necessary to mitigate the negative impact of a free-market system in Earth orbit. However, existing space treaties do not provide explicit mandates for orbital debris removal, nor is it clear how such regulations would be upheld. But yet space debris is an untapped resource in itself.

There is currently more than 13,000 tonnes of junk in Earth orbit. By comparison, the ISS weighs 450 tonnes. Space 'debris' typically consists of components that could be repaired or harvested for reuse or contain rare metals, all of which have been extracted, processed and launched from Earth in resource-intensive processes. There are proposals under development to remove debris from Earth orbit by retrieving old equipment and fragments thereof via magnetic effects, harpoons, claw-like structures or sticky robotic arms, then typically pushing them towards Earth to burn up in the atmosphere. From a resource efficiency perspective, can we do better?

Automated and crewed on-orbit servicing infrastructures could provide maintenance, refuelling or repair services for failed satellites in orbit; furthermore, component- and material-recycling coupled with 3D printing technologies could enable the production of satellites in orbit. Manufacturing in space is within reach, enabled by developments including the increasing miniaturisation and modularisation of many satellite technology components; developments in additive-manufacturing, including the demonstration of 3D printing in the ISS; as well as impressive advancements in robotics and automation, particularly on the surface of Mars.

Beyond the studies typically performed by just six people in the ISS, with bigger teams and more facilities, research, innovation and production in the microgravity environment of space could be extended to include the development of novel propulsion systems, new materials, modular life-support systems to be serviced and maintained in space, robotic manufacturing and assembly, spacecraft design, and off-world supercomputers to support this activity, to name but a few.

Our activities in Earth orbit have thus far provided a good precedent for international cooperation among crews representing a (small) range of countries. While nearly half of the countries on Earth have their own satellites in orbit, just a few have access to the research and development, and more recently tourism, taking place in the microgravity environment of Earth orbit through human presence there. Canada, the eleven member states of the European Space Agency, Japan, Russia, the US and more recently, with their own space station, China currently fund and access facilities in orbit; and with the rise of space tourism, more citizens who are wealthy enough to afford it, mostly from these same countries, are joining in.

We've been launching things into space since the 1940s. We have been living in space continuously for the past twenty years. What's next? It's time we advance to being able to produce space technologies in space, from resources we extract there. To utilise space debris we will need more diverse crews as well as robotic support in orbit, which will require more space station facilities. At a relatively safe distance of a few hundred kilometres above home, where resupply and evacuation are just hours away, more people living and working in the microgravity environment of Earth orbit will provide crucial data for the teams and the technology that we will need to explore further beyond Earth in the coming years.

We have the tools at our disposal to develop the infrastructure required for our sustainable use of Earth orbit. Assuming we want to continue our exploration of the space beyond our planet, we need to reflect on whether the economic and regulatory systems that drive and manage activities in Earth orbit are aligned with this objective. Our very ability to leave the surface of the Earth in the future depends on our ability to successfully manage the space above our planet.

It seems unlikely that new economics, prioritising efficient resource utilisation or exploration ahead of profits, will emerge as a result of activities in such close proximity to Earth. So for now, we can only hope that the commercial value of data obtained in Earth orbit and the potentially reduced costs of repurposing space debris for satellite manufacture, coupled with revenue from space technology innovation in orbit, will suffice to support our continued activity there. Just as the view from Earth orbit enables revolutionary new ways to understand our home planet, for example in climate monitoring, perhaps the perspective from the Moon, and the in situ resource utilisation and manufacturing required for a permanent human presence there, will inspire the desire and the capability to take better care of Earth orbit.

Men on the Moon

This is the decade when we return to the Moon! It has thus far been underutilised, with only four American men still alive today who have walked there (for the last time in 1972); but plans to return to the lunar surface and build habitable bases there are finally being formulated by China, the US and their various partner states. A preliminary crewed lunar orbital mission scheduled for 2025 will be the first time we have gone beyond Earth orbit in over a half-century.

Our natural satellite, the Moon, is on average 385,000

kilometres – or just three days – away. It is also one of the harshest environments in the Solar System, with temperatures ranging from negative 250 to over 120 degrees Celsius during two-week-long days and nights, no atmosphere to speak of, surface radiation over 100 times greater than and gravity just 17 percent of Earth. Our Moon is therefore a conveniently nearby resource for research, innovation and the exploration of more distant off-world destinations.

The first Moon landings required further sophistication of life-support systems than in Earth orbit, with crews being much further away for just over a week. The Apollo spacecraft had three parts: a command module including a cabin in which the crew would return to Earth; a service module providing propulsion, power, oxygen and water; and a lunar module consisting of a lander and an ascent stage to get the astronauts back into lunar orbit. The Apollo crews had three outfits: a cotton all-in-one undergarment; an inflight non-flammable Teflon jacket, trousers and boots; and finally a spacesuit, worn for launch, re-entry and on the lunar surface. During the journey, the crew breathed normally as the spaceship was pressurised with an on-board oxygen source. Food and drink came freeze-dried and had to be rehydrated by adding water. Waste, including human waste, was put in bags with germicide pills to prevent fermentation and gas production, and then stored in a waste disposal compartment.

The first humans to walk on the Moon, astronauts

Neil Armstrong and Buzz Aldrin, spent almost twenty-two hours on its surface, during which time pilot Michael Collins circled the Moon, completely alone, once every two hours. The first layer of the spacesuits worn on the surface was water-cooled underwear, next a five-layer, airtight pressure garment, then a thirteen-part outer garment to protect the men from micrometeorites, ultraviolet light and other radiation. The suit could sustain activity for up to four hours before requiring recharge. The portable life-support system was worn like a backpack over the suit. Its functions included regulating suit pressure; providing breathable oxygen; removing carbon dioxide, humidity, odours and contaminants from breathing oxygen; cooling and recirculating oxygen and water; two-way voice communication; and a display of suit health parameters and wearer's health, including heart rate.

While a revolution in propulsion systems is not necessary to get to the Moon and back, human presence there will require resources. The equipment needed for shelter, power, mobility, communications and to produce oxygen and water on the surface will initially come from Earth; the ability to maintain, repair or augment this infrastructure locally with lunar resources is an active area of investigation.

Our data and samples show that the elemental composition of the lunar surface is dominated by oxygen and silicon, giving weight to theories that it was once part of

Earth's crust; however, there's more iron and titanium, while carbon, nitrogen and alkali metals like sodium and potassium are less abundant. Our knowledge of what lies below the surface is limited; the Kaguya spacecraft, launched in 2007, detected the presence of the heavier radioactive elements uranium and thorium in lunar dirt. Helium-3 is thought to occur in greater abundance on the Moon than on Earth, and is an appealing candidate fuel for nuclear fusion, which could be commercialised in the coming decade.

Critical for establishing a human presence is our detection of water ice in the permanently shadowed craters on the Moon's surface. Water supports life, in liquid form as well as from its oxygen content. Through the process of electrolysis, the hydrogen and oxygen atoms in water can be split apart, providing fuel in the form of hydrogen, which can then be burned in oxygen. Decades of innovation in air and water management as well as radiation protection for the crews in Earth orbit will provide an important basis for the life-support systems deployed on the Moon. With the technological advances that have taken place since we were last there, grand visions for infrastructure on the Moon are realisable.

If we get it right, human presence on the Moon will be an opportunity for collaboration-driven innovation in an extreme environment, a laboratory for space exploration in relative proximity to Earth and a stepping-stone to Mars and beyond. The data we collect from wearables, surface exploration, subsurface investigation and the performance of innovations in life-support systems increasingly fuelled by in situ resources will enable us to prepare for establishing off-world communities throughout the Solar System.

The nearside of the Moon is always in line of sight, with just over one second's communication delay to Earth. Beyond the Earth–Moon system however, for example on Mars, the situation will be quite different. But let's not rush ahead; over half a century of experience in Earth orbit has generated a wealth of knowledge and inspired humans all over the planet to look up, knowing that there are precious humans inside the space stations sometimes visible hurtling across the night sky.

While we have not yet returned to the surface of the Moon in fifty years, in 2023 the Chandrayaan-3 mission landed for the first time at the lunar south pole, setting the stage for future sample-return missions. More recently, in early 2024, Japan became the fifth country to execute a soft landing on the Moon and, using facial recognition algorithms to identify craters, is the most accurate landing yet, at less than 100 metres from the target site.

Just a handful of missions so far have been successful at surviving the lunar night, which plunges to temperatures far below those experienced anywhere (outside the laboratory) on Earth and even in low Earth orbit, where rapid orbital speeds enable regular solar warming, such that satellites are designed to withstand temperatures down to about negative 65 degrees Celsius. By our definitions of temperature, negative 273 degrees Celsius is absolute zero – the coldest anything in the Universe can get, theoretically anyway – so the Russian, Chinese and Japanese missions that have been successful in surviving the lunar night are important demonstrators of technology functionality in such cryogenic conditions – and important precursors to the plans to establish life on the Moon which, unlike the Apollo missions, are planned for longer than the fourteen (Earth) days of lunar sunshine.

Bases on the Moon could include modular shelters with panels produced from lunar regolith; power production redundancy through periodically abundant solar energy, nuclear power, as well as hydrogen fuel cells; wastewater-processing to output hydrogen as well as nutrients for lunar food production; indoor agriculture and cultured meats from stem cells; solar- and hydrogen-powered pressurised rovers; the list goes on. Besides extracting lunar ice for water, hydrogen and oxygen, or uranium or thorium as fuel for nuclear devices, proposals to mine metals like titanium and silicon from the surface of the Moon and vacuum-deposit them to produce solar cell or even telescope components locally are also under consideration.

At just over a light second away, the nearside of the Moon is close enough to consider integration with the existing Internet. Connectivity on the farside will require lunar satellites, which can also be deployed for signal transmission between distant locations on the surface. There are currently a handful of active lunar orbiters, launched by China, India, Japan, South Korea and the US. Future lunar communication satellites and local networks, computing power, surface mobility vehicles, healthcare facilities and so on could become an opportunity for joint use and collaboration between the nations and private companies active there.

Lunar cryptocurrencies and locally implemented cryptocurrency-mining could form part of the lunar economy, facilitate trade between the Moon's outposts and provide the financial infrastructure for lunar-based transnational projects to trade resources or information with organisations back on Earth.

So who on Earth will pay for all of this? Of late, as inequality increases, our economic system is producing greater numbers of billionaires, wielding net worths exceeding many national space budgets. A new era of public–private partnership in the space industry is stimulating activity the likes of which we haven't seen since the Cold War. Commercial resource utilisation is central to the US endeavour to return to the Moon; NASA has announced that it will buy resources such as ice, rocks, dirt and other lunar materials to stimulate private extraction of off-world resources for use in space. China, on the other hand, is making rapid progress towards the utilisation of space resources towards its own national economic development.

Just how profitable could lunar resources be? In the

past few years, launch costs from Earth have dropped significantly, and while the fully reusable Starship may see this reduced by orders of magnitude yet, it still costs over 1 million US dollars to get a tonne to orbit. The value of extracting water on the Moon, whether for local use or for refuelling within the Earth–Moon system, will be determined in relation to such costs. Beyond water, here on Earth a range of factors from environmental to societal can result in shortages in metal and mineral supplies, and disruptions in supply chains. For example, the economic lockdowns of the COVID-19 pandemic, combined with severe weather, played a central role in the global semiconductor chip shortage crisis; and more recently, conflict in the Middle East has resulted in shipping disruption through the Suez Canal, all having a global impact on availability of goods and resources. It is not impossible that future shortages caused by a combination of international conflict, sanctions, extreme weather, a consumerist culture and ongoing shortcomings in waste management could result in it becoming cost-efficient to deliver certain resources or products from the Moon to the Earth.

Besides mining or manufacturing, there is commercial opportunity for the information produced on the Moon; from documentary footage to research and technology development data. Private companies are also evolving capabilities to take people to the Moon, raising questions about how restrictions on tourist activities on the lunar surface could be enforced.

Similarly to Antarctica, activity on the Moon is governed by international treaties. With line-of-sight communication and accessibility all year round, the Moon is arguably less remote than Antarctica. Another difference is that commercial activities in Antarctica are in general explicitly not permitted, and perhaps as a result the continent has never seen large-scale conflict. Once it has been established that there is money to be made on the Moon, will regulation keep up? The unchecked accumulation of debris in Earth orbit as the result of an emerging trillion-dollar industry doesn't bode well.

The Moon Treaty of 1979 disallows states to conduct commercial mining on celestial bodies until there is an international regime for such exploitation; however, it has not been ratified by any nation able to launch crews. The Outer Space Treaty of 1967, on the other hand, has been signed by most countries on Earth and limits the use of the Moon and other celestial bodies to peaceful purposes for exploration and utilisation by all nations. But it is largely silent or ambiguous on commercial activities on the Moon, including resource extraction such as lunar mining. Whether national or private, when large profits are at stake international agreements can be de-prioritised and the outcome can be conflict.

While we haven't been beyond Earth orbit in the past half-century, particularly impressive improvements in remote exploration and the automation of systems to collect data in off-world locations promise to enable exciting progress for crewed exploration in the near future. On Earth, our unprecedented and rapid technological development has seen the continued growth of the human population as well as life expectancy increase in many regions; as a result, the largest-ever population of humans with the greatest-ever set of resource requirements are producing unprecedented volumes of data. These developments are sufficient to equip us for our continued expansion beyond Earth, but can they save us from ourselves? While space may be considered an extreme environment, the conditions in most parts of the Solar System are relatively predictable by physics and chemistry. The biology here on Earth – and in particular a population of over 8 billion humans vying for limited resources, with our activities having an increasingly destructive impact on the biosphere – makes our home planet a comparatively unpredictable place to be. While this may drive accelerated progress off-world in the coming years, one thing we can predict for sure is that things on Earth are not going to get less extreme anytime soon.

6

WORLDSHAPERS

As the saying goes, the only constant is change. Life requires resources, living systems access these resources from their environment, and the constant and often dramatic changes in the conditions on this planet over the past 4 billion years have given rise to the awe-inspiring complexity that makes up our terrestrial biosphere. Among that complexity is us: curious explorers who have shown remarkable resilience to great environmental shifts in the past through our ability to migrate, our use of tools, social cooperation and eventually through the development of the capability to shape the environment around us. However, this ability can have – has had – unintended consequences, in particular when coupled with the historically prevalent belief that we are separate to the natural world, have dominion over it, and may therefore plunder it with impunity.

Recalling the Milanković cycles of the distribution and the seasonal cycle of solar radiation across the planet as outlined previously, we may ask where we are now with respect to the orbital factors that impact our climate.

We are currently approaching our most circular orbit in the cycle, a time when the length of each of our seasons is about equal; precession is making southern summers hotter and moderating seasonal variations in the north; while our obliquity is currently about halfway between its extremes, decreasing to its minimum over the next 10,000 years, a period well correlated with glaciation in Earth's recent history. All of this points to a current cooling trend, estimated to have begun some 6,000 years ago, which should extend several thousand years into the future. However, this is not what we are observing.

The atmosphere of our planet – a mixture of gases that extends from the land and sea to the edge of space around 100 kilometres up – plays an important role in determining the conditions near the surface, where most life resides. Owing to its gaseous state, changes in atmospheric conditions, what we call weather, can impact large regions in short timeframes, compared with changes in the crust or oceans of the planet. Climate refers to the prevailing weather conditions over long periods.

By examining the fossil record, we see a range of rapid, large-scale decreases in global biodiversity – what we call mass or global extinction events – punctuating the history of life on our planet. Such extinction events typically occur when the environment changes more quickly than species can adapt, so that all members of a range of species are unable to procure sufficient resources to survive in the new conditions. Extinction events are often correlated with phenomena like impacts with space rocks or volcanic eruptions, which cause a sudden injection of gases into the atmosphere. This large-scale disruption of atmospheric conditions can have a significant impact on the global climate. Vast activity by life forms themselves can also alter the atmosphere on a planetary scale: cyanobacteria triggered a major extinction event when they evolved water-splitting capabilities to produce fuel and began to oxygenate the atmosphere over 2 billion years ago. Recently, we have also started to cause global changes to atmospheric conditions. Our species has a history of shaping our global environment, though never on a larger scale than now.

Legacy power systems

Without energy, nothing much happens. Our recent accelerated technological development and population growth are coupled with an insatiable need for this most fundamental of resources. From an estimated few thousand terawatt (trillion-watt) hours obtained from burning biomass in 1800, today global energy consumption has increased over thirty times, the bulk of it, however, still being obtained from burning organics. And we are going to continue to need more, a lot more if we are to evolve to a Kardashev level-one civilisation that uses an amount of energy equivalent to that produced by its Sun. The current climate crisis, however, has been driven primarily by our current methods of generating power, rather than by volume.

The climate is not the only crisis emerging from our power-production methods; our society has been forged on the burning of finite fossil fuel stores, and competition to control the remaining deposits continues to result in conflict, displaced people and death in many parts of the world. Surely our power-hungry technologies and the science on which they're built can also provide novel ways of thinking about power generation?

Around a million years ago, our early ancestors learned to control fire, extracting energy from the dead biomass produced by photosynthetic life forms. Next, we discovered fossil fuel: material formed over millions of years from the fossilised remains of these organisms by exposure to heat and pressure beneath the Earth's surface. We discovered that burning these fuels produces a lot more energy than burning, for example, wood.

The subsequent, and rather rapid (that is, compared with the technological evolution of the previous million years) invention of engines, first driven by coal and steam and then by internal combustion, culminated in the Industrial Revolution of the eighteenth century in Britain, Europe and the US. This era saw a transition to mass-production of goods using recently invented engine machinery, and was facilitated by the new economic concept of manual labour – long, repetitive hours of human work – as well as by the colonisation of various other parts of the world for the required raw materials. People began to move to cities to get paying jobs in industry in order to buy the goods being produced, and public transportation, communication and banking systems emerged. Thus, with manufacturing, the cogs of capitalism began to turn, propelled by the burning of fuel. However, alternatives to combustion to produce electricity were already being developed during this time; fossil fuel alternatives might feel like a twenty-first-century invention, but their beginnings are centuries old.

In the early 1800s, inventors Humphry Davy and William Grove conceived of simple hydrogen fuel cells; by combining hydrogen with oxygen, electricity is produced with only pure water as a by-product. In the 1880s, Charles Fritts built the first commercial solar panels, with a thin layer of selenium and an even thinner layer of gold, resulting in a solar to electrical conversion efficiency of about a percent. The potential of such developments was huge, but only much more recently have these kinds of power-production technologies entered the mainstream, largely in the context of the automotive industry. In fact, transportation systems have played and continue to play a critical role in driving our methods of producing power.

In 1913, Henry Ford implemented the first moving assembly line for the mass-production of an entire automobile, pioneering automation in manufacturing, reducing the time it took to build a car from over twelve hours to one hour and thirty-three minutes and revolutionising the way people moved around. Early cars could run on a variety of fuels including steam, electricity (and we already had a variety of methods to produce it) and ethanol. Unlike gasoline, which is produced from fossil fuel crude oil, ethanol – a biofuel – can be produced from agricultural waste, is therefore renewable, and its usage emits up to 50 percent less greenhouse gases. Ford realised the potential of biofuel all those years ago, calling it 'the fuel of the future'. But by 1920, 9 million gasoline-powered vehicles were on the road in the US.

It seems that other factors besides logic were at play, leading to oil supremacy in the vast majority of our journeys today. Surely not coincidentally, around that time oil magnate John Rockefeller, who at one point controlled around 90 percent of all oil in the US, became the first person ever to acquire a personal fortune of a billion US dollars (roughly the equivalent of 445 billion today – a net worth more than that of anyone alive today), and is widely regarded as the richest person in modern history. To power our new machines, and in spite of the alternative power-production strategies at our fingertips, we burned and burned the stuff, arguably resulting in much of the technological development that we enjoy today, but also in the pollution of our air and an atmospheric composition that traps more heat from our Sun.

Today almost 100 million cars are produced annually around the globe, and roughly 100,000 passenger flights take off daily, powered largely by a trillion-dollar global fossil fuel industry. Mass-production and fossil fuel have transformed the way we travel. While hugely beneficial to a few, the choice to power global transportation systems with fossil fuels remains a major contributor to pollution, disease and global warming, contributing to millions of premature deaths annually and almost 30 percent of global carbon emissions.

Access to energy is fundamental to much of the functionality of any human community. Just ask someone living in South Africa about the strategies we employ to produce our own power while experiencing load-shedding, or scheduled rolling blackouts, for up to six hours at a time over the past twenty years, or the almost one billion people on Earth who still don't have access to an electrical grid at all. Our power-production strategies, or lack thereof, are a major contributing factor to the global climate shift we are experiencing here on Earth. Needless to say, the future – both on- and off-world – will require unseating the legacy of the combustion-power-production techniques that have dominated our society for the past couple of centuries. Not only are they inefficient but they are harmful to life. To do this, we must first go back to basics.

A fundamental requirement for life, or really for anything interesting, is energy. Over the past 4 billion years, photosynthetic life on Earth has absorbed the energy streaming in from our Sun and stored it in the chemical bonds of organic compounds. These are a large class of chemical compounds in which atoms of carbon are linked to atoms of, most commonly, hydrogen, oxygen or nitrogen. Around a million years ago we harnessed the power of fire, extracting energy from dead photosynthetic organisms like plants and trees. Next, we discovered fossil fuel, which can be burned in our unique oxygen-rich atmosphere to release even more energy to provide heat, power engines or generate electricity.

Here on Earth, burning photosynthetic biomass is an indirect means of harnessing the continuous stream of energy being emitted from the Sun; this is not a strategy to be relied on beyond our home planet. The photovoltaic effect, discovered in 1839, is the process of generating electric current from sunlight. A solar cell is an electronic device that performs this task. In certain materials, the absorption of light transfers energy to electrons, resulting in the movement of these higher-energy electrons, which can then generate a current. Semiconductors have this property, and typically the abundant semiconductor silicon is employed in solar cells. For efficient photovoltaic operation, solar-grade silicon needs to be 99.9999 percent pure. Pure silicon is derived from abundant and naturally occurring minerals like silicon dioxides, for example crushed quartz, and a series of non-trivial heating procedures are employed to remove impurities. The pure silicon is then doped to enable conductivity, and an anti-reflective coating is added to the surface to increase light absorption. Metals are used to conduct current out of the semiconductor material into an external circuit, as well as being employed for enhanced efficiency. These include metals that are toxic and relatively rare, or both, such as cadmium, gallium, indium and lead. While thin-film, lightweight and flexible solar cells such as amorphous silicon cells are improving in efficiency (to around 50 percent solar conversion in laboratory conditions), on any celestial body with a day–night cycle or weather we still need to consider storage requirements for this energy.

Typically, batteries are employed to store solar energy until it becomes available again. Chemical batteries are devices that store energy chemically and then convert it into electricity. The three basic components of a chemical battery are a metal that tends to lose electrons, a metal that tends to gain electrons, and a solution between the two that can conduct electricity between them. This is how we have been making batteries for the past few hundred years; or maybe even more. The Baghdad Battery, believed to be about 2,000 years old, consists of a ceramic pot, a tube of copper and a rod of iron. Tests of the inside of the pot show that it contained something acidic like wine or vinegar. Sadly, and not without irony, the Baghdad Battery went missing after the US invasion of Iraq in pursuit of the control of extensive oil deposits found there.

Batteries today may be more efficient than ever before, but typically require intensive manufacturing procedures as well as, just like solar cells, depending on a range of rare and often toxic metal resources whose extraction, transportation and processing is energy-intensive and has a detrimental impact on our society and environment. Furthermore, batteries remain relatively large considering their limited lifetimes, and in the US alone an estimated 3 billion of them are thrown away each year, generating vast amounts of hazardous waste.

Can we do better? As we develop the experimental techniques to probe life's oldest process on small scales, we learn more about how photosynthetic organisms store sunlight. Within the first nanosecond of absorbing a photon of sunlight, a photosynthetic complex converts this energy into the movement of an electron from one place to another, generating a charge separation between the positive 'hole' that is left behind and the negative charge of the electron at its new position. This charge-separated state is an effective battery that powers subsequent processes in the cell. This process happens at near-perfect efficiency of 100 percent. While bio-inspired solar energy-harvesting and storage systems are an active and exciting area of research, our incomplete understanding of biology from a physical and chemical perspective has not yet produced a fully realised area of artificial photosynthetic technologies.

But, looking further, photosynthesis doesn't stop at the production of a charge-separated state. Plants, algae and bacteria are doing more than just harvesting sunlight; they are employing protective mechanisms and, crucially for this discussion, continuously building biomass and producing fuel such as glucose.

For our purposes, analogously, rather than store energy in batteries, we could use sunlight energy directly to produce biofuel, whether from organic waste from agricultural or human activities or biomass grown for purpose. Brazil has already embarked on such a programme, utilising primarily sugar cane and its by-products to produce ethanol which constitutes up to 25 percent of its standard petrol blend and contributes more than 5 percent of the country's total power consumption to the grid.

While replacing the fuel on which existing infrastructure, from cars to power plants, runs with biofuel is not a bad idea on Earth as we face the eventual depletion of fossil fuel, why limit ourselves to the centuries-old legacy of combustion engines? Instead of generating heat and then electricity, we can produce electricity directly from fuel. A fuel cell is a device that converts the chemical energy in a fuel into electricity through a chemical reaction rather than by combustion. Fuel cells do not need to be recharged like batteries, but instead continue to produce electricity as long as a fuel source is provided. A biological or a microbial fuel cell converts chemical energy into electricity by the action of microorganisms. A range of organic materials can be used to feed the fuel cell, including wastewater. And the output? Cleaner water.

The simplest fuel cell, invented centuries ago, also produces water as a by-product: the hydrogen fuel cell. Here, hydrogen and oxygen are combined to generate electricity, heat and water. As there are no moving parts, fuel cells operate silently and with high reliability. Hydrogen is the simplest and most abundant element in the Universe, and the by-products of generating electricity, namely heat and water, are valuable resources in themselves. And while water ice is abundant throughout the Solar System, on Earth we have hydrogen on tap from the oceans that cover 71 percent of our planet. Is the hydrogen fuel cell the ultimate power source? Archaea have been using hydrogen as a source of energy for billions of years, inspiring new methods of its industrial production by the complex enzymes they employ for its manufacture and use. But, in fact, we can go even deeper.

The most fundamental strategy of power production that we know of, millions of times more efficient than coal, oil or gas, is via accessing the energy present in the nuclei of atoms. Just ask the Sun. Inside stars, the light nuclei of hydrogen are fused into helium, thereby releasing power. Exciting developments towards controlled and scalable nuclear fusion are continuously taking place; the main challenge is to reach the extremely high temperatures for long enough to make fusion happen. Since reactor pressures are much lower than in the Sun, temperatures of at least 100 million degrees Celsius are required: around seven times hotter than the core of the Sun. Recently, a US facility has announced the first ever breakeven experiment, where fusion energy surpassed the energy delivered, while in South Korea, the fusion research device called the 'artificial sun' sustained plasma at such temperatures for almost a minute. Some estimates anticipate clean and safe (there are no greenhouse gas emissions or radioactive waste) commercial electricity from nuclear fusion by 2030 – perhaps just in time for the planned off-world settlements of the near future, particularly on the Moon, where the quantity of helium-3 (an isotope of helium which could be used as fuel for future nuclear fusion reactors) is thought to be greater than in places with an atmosphere like Earth or on Mars.

Rather than creating an environment where nuclei can be fused, we can also look to naturally occurring radioactive elements that decay spontaneously, emitting radiation. On Earth, the Moon and likely most bodies in the Solar System, the primary sources of natural radiation are from the heavy elements uranium and thorium. These naturally occurring decays warm the Earth, contributing to the maintenance of our liquid core that sustains the magnetic field, which in turn supports terrestrial life. Nuclear fission reactions entail the subdivision of heavy nuclei accompanied by the release of a large amount of energy, typically a few times less than by fusion. The majority of electricity from nuclear power is currently produced by the nuclear fission of uranium; specifically, the uranium-235 isotope. These emissions have the potential to kill cells in living organisms, or cause mutations to DNA, which can cause cancer, and shielding this radiation is thus an important aspect of producing nuclear power. Barriers of lead, concrete or water are typically employed. A by-product of nuclear fission is radioactive material, also called nuclear waste, which cannot be destroyed by any known process. On

Earth, what can't be recycled is disposed of in canisters, which are placed in tunnels, which are subsequently sealed with rocks and clay in underground repositories.

While both fission and fusion power generation are literally millions of times more efficient than the burning of fossil fuels in terms of the amount of energy obtained from a given mass of fuel, many of us remain anxious about nuclear power. In reality, on Earth far more people die from air pollution due to the burning of fossil fuels than have ever died in nuclear reactor accidents. The total deaths since the 1950s from nuclear accidents number far fewer than a million by any counts – some claim numbers are as low as a few hundred – while around 7 million people die prematurely from air pollution every year.

Where does this fear come from? Well, maybe because analysis shows that the detonation of just 100 nuclear weapons could end much of life on Earth, in particular complex life like humans. This is coupled with the inherited fear struck into the hearts of everyone alive in 1945 when as many as 200,000 citizens were instantly annihilated, a similar number dying subsequently of radiation poisoning, when the US deployed nuclear weapons without warning on the people of Japan. Around half of the populations of Hiroshima and Nagasaki were wiped out; exact numbers were difficult to tally in the aftermath of fire, rubble and chaos. Blast shadows in the museum in Hiroshima that were once human beings going about their daily lives – their outer bodies carbonised by the thousands-of-degree-Celsius heat of the explosion before they burned in the subsequent 3-kilometre-wide firestorm – are permanently imprinted into the brick wall, as they are forever etched in my memory. At the time I lived in Japan, cocktail-bartending for the Yakuza; but that's a tale for another time.

So what is the connection between nuclear power generation and the production of nuclear weapons? Let's go back to the two candidates for producing energy by nuclear fission: uranium and thorium. Why haven't we heard much about thorium? Thorium, named after Thor, the god of thunder in Norse mythology, is a radioactive metal found in igneous rocks and sand. In fact, certainly on Earth anyway, compared with uranium thorium is more abundant; thorium-based fission is more efficient and produces less waste and fewer long-lived radioactive isotopes than conventional nuclear reactors; and furthermore, the possibility of a reactor meltdown can be eliminated. So why on Earth aren't we using thorium? The answer is based in war-time politics: uranium-based fission produces plutonium as a by-product, which is the primary fuel used in nuclear weapons. The by-products of thorium, on the other hand, are much more difficult to weaponise.

Weaponisation, or the difficulty thereof, aside, thorium has a lot of potential for power generation on Earth and beyond. Thorium-232, the only naturally occurring isotope of thorium, has a half-life of 14 billion years – the estimated age of the Universe. Applying high-energy neutrons can cause it to undergo fission, eventually forming uranium-233, a fuel that can be used in nuclear reactors. One candidate for thorium-based nuclear power production is the molten salt reactor, which contains a hot liquid salt in which a nuclear reaction takes place. The salt consists of the nuclear fuel and some other compounds which optimise the reaction, the heat transfer and the stability of the salt. The good heat transfer capacity of the molten salt allows for quick transfer of a lot of energy from a relatively small reactor core. The salt mixture is both the fuel and the coolant. This allows for high operating temperatures, efficient electricity generation and the potential utilisation of the heat for other applications, while not requiring water for cooling or pressurisation for functionality, which eliminates any risk of explosion. Also, nuclear meltdown is not possible as the molten salt is already in liquid form; any kind of loss of control of the reaction results in automatic shutdown of the system.

While molten salt thorium reactors remain developmental, some estimates indicate that one tonne of thorium can produce as much energy as 200 tonnes of uranium, or 3.5 million tonnes of coal. India has some of the world's largest thorium deposits, and is aiming to produce 30 percent of its power from thorium fission by 2050, while China has recently announced plans to establish the world's first thorium molten salt nuclear power station in the Gobi Desert in 2025.

For smaller-scale power requirements, nuclear

batteries can be classified by energy conversion technology into two main groups: thermal and non-thermal converters. The thermal types convert some of the heat generated by the nuclear decay into electricity. The most notable example is the radioisotope thermoelectric generator, often used in spacecraft like the Voyager missions nearing a half-century of operation. The non-thermal converters extract energy directly from the emitted radiation before it is degraded into heat. They are easier to miniaturise and do not require a thermal gradient to operate, so they are suitable for use in small-scale applications. The most notable example is the betavoltaic cell, recently announced to be nearing commercialisation, for applications requiring up to a watt of power (approaching the average consumption of smartphones of around 5 watts), by a company in China.

Our science and technology have revealed a plethora of ways to produce power for society's growing needs. But our history of power production has been motivated by greed and struggles for power rather than logic or efficiency. And these behaviours are having consequences.

The Anthropocene

We have an era named after us; not because of our technological prowess, but because of the destructive impact our industry is having on our planetary biosphere. The Anthropocene Epoch, also referred to as the Anthropocene Extinction, is a unit of geological time describing the most recent period in Earth's history; a time when human activity has begun to have a significant impact on the planet's climate and ecosystems. Our current way of doing things can only get us so far; there are numerous indications that we are close to that point.

In the past 200 years, we have evolved the capability to harness the sunlight energy stored in ancient biomass: fossil fuels. Our vast industrial activities powered by burning these fuels are resulting in the increased prevalence of, primarily, carbon dioxide (CO_2), but also methane and other greenhouse gases in our atmosphere. Air bubbles trapped inside ice core samples, which in Antarctica date back almost a million years, are our oldest direct evidence of CO_2 levels. However, by studying things in the geological record known to be correlated with CO_2 levels, for example strontium-isotope records, we can estimate that they may have been as high as 4,000 parts per million around 500 million years ago. The Antarctic ice cores indicate that over the past 800,000 years, CO_2 levels have remained between 170 to 300 parts per million. For the past 200,000 years that *Homo sapiens* has been around, we see a maximum of nearly 300 parts per million during the interglacial period that began around 130,000 years ago; but by 2013, CO_2 levels surpassed 400 parts per million for the first time, not only in recorded history but also in the entire 2 million years that our genus *Homo* has walked the Earth. Levels continue to rise, and at this rate we will reach levels of 600 parts per million, double the maximum during our existence, by 2050.

But what is wrong with CO_2? Earth has experienced higher levels than those prevailing today, plants use it, and with carbon and oxygen in the top five most abundant elements in our Galaxy, it's not an uncommon and certainly not an unnatural molecule. However, unlike the cyanobacteria which triggered the first major extinction event on Earth by the emission of oxygen, we are aware of the potential implications of what we are doing. We have climate data on much of the history of life on Earth, and we know that 99 percent of species that have ever lived on this planet have gone extinct. We have seen that sudden climate changes have the potential to wipe out up to 96 percent of all extant species, and that often the more complex creatures at the top of the food chain are the first to go. We also know that beautiful sophistication can emerge from major climate changes; like multicellular life, including all plants and animals, for example. Life, in some forms, will prevail. The question we need to consider is: will we be a part of the next wave of complexity to emerge from the Anthropocene Extinction or not?

A greenhouse gas absorbs heat radiation from the Sun and then traps it in the Earth's atmosphere. On a planetary scale, whether climate conditions are undergoing heating or cooling is determined by how much of the Sun's energy the Earth absorbs compared with how much it radiates through the emission of infrared heat. Increased amounts of greenhouse gases in the atmosphere cause warming on the surface of the Earth. The majority of this warms the ocean, with the remainder heating the land, melting snow and ice and warming the atmosphere. When we compare satellite observations of the total energy absorbed by Earth with the amount emitted back into space, we see that lately Earth is trapping more than double the amount of heat that it did in 2005. Independent measurements of heating within the ocean, land and atmosphere, and melting of snow and ice, confirm these results.

We've heard about exponential population growth; the shortest period over which our population has doubled is thirty-seven years in just over a generation, from 2.5 billion in 1950 to 5 billion in 1986. Anything doubling on a planetary scale in a single generation is something to pay attention to. Again, Earth is retaining more than double the sunlight heat than it did just twenty years ago. We are already seeing the impact of this in an increase in extreme weather events, including heatwaves, droughts, flooding and tropical storms.

Carbon dioxide is not the only thing we are putting into the air. Air pollutants in general – including carbon monoxide, nitrogen dioxide, sulphur dioxide and lead, all emitted by the burning of fossil fuels – can poison living organisms through the disruption of endocrine function, organ injury, increased vulnerability to stresses and diseases, lower reproductive success and possible death. The combined effects of atmospheric and indoor air pollution are associated with millions of deaths each year, dwarfing reported COVID-19 mortality rates by comparison, and making it one of the top contributors to human demise. More than 90 percent of people on Earth are estimated to breathe polluted air.

Atmospheric disturbances are soon transferred to our planet's water. Covering over 70 percent of our planet's surface, our marine biosphere is currently threatened by overfishing, pollution and climate change. It is estimated that by 2050 there could be more plastic in the sea than fish by mass; some say that this tipping point has already occurred. Owing to industrialised fishing, today's seas contain only 10 percent of the marlin, tuna, sharks and other large predators that were found in the 1950s. In addition, with increased air temperature and carbon dioxide levels, seawater holds less oxygen and becomes more acidic, resulting in a reduced ability to support life. We see significant fossil size decrease with increased sea temperatures in the past. Ocean acidification can dissolve the skeletons of krill and plankton on which many animals, including whales, rely as a food source; recall the extinction event the Great Dying? When I met oceanographer Sylvia Earle (also known as Her Deepness) a few years back at dinner at an event in Lindau, a student asked where on Earth her favourite diving spots are. She replied, 'Anywhere, fifty years ago.' A sobering thought. When the waiter then asked if she wanted red or white wine, she said both, and I immediately followed suit.

Our planet is unique in the Solar System in having liquid water on the surface; all known living organisms depend on this precious resource for survival. However, even though liquid water covers over 70 percent of our planet's surface, and in spite of our great technological 'progress', water shortages may be the most dire challenge we will face in the coming decades. Only around 3 percent of water on Earth is freshwater, the vast majority of which is either stored as ice or deep underground. Human activity is putting these freshwater resources under stress and our aquifers, rivers and lakes are becoming too polluted to use or are drying up.

Besides drinking, washing, sanitation and agriculture, much of the water we consume as a society is used in industry: it takes around 20,000 litres of water to produce a kilogram of chocolate, while the total water footprint of soft drinks like cola is allegedly hundreds of litres of freshwater for just one litre of soda, claims denied by the top corporate culprits. Drought already impacts 40 percent of our population, with nearly two-thirds of humans facing water shortages in at least one month of the year. The gap between global water supply and our societal requirements is projected to reach 40 percent by 2030, with over three-quarters of the world's population predicted to be affected by water shortages by 2050.

Climate change is contributing to water scarcity by altering weather and rainfall patterns around the world, resulting in greater prevalence of drought and increased rates of glaciers, ice and snow melting into the ocean. Agriculture currently accounts for more than half of water usage globally, and with the increase in food production for our burgeoning population, the demand for freshwater will also grow. Conflict over water resources is already taking place: the worst drought on record in the Horn of Africa in the early 2000s is one of the reasons for the ongoing conflict there. The collapse of agriculture, urban migration, uprisings and ensuing civil war are contributing causes of the world's largest refugee crisis, with over 12 million Syrians displaced from their homes to date. By 2030, some estimates predict a billion climate refugees.

Soils simultaneously purify water, produce food and store carbon. Vast regions of the Earth's surface have been disrupted by human activities; since the Industrial Revolution, over 100 billion tonnes of soil have been lost from farmland. Microbes, including fungi and tiny worms called nematodes, turn waste into nutrients in the soil – Nature knows no waste – working symbiotically to help trees and plants grow. But our agriculture and use of fertilisers, pesticides and antibiotics, as well as our destruction of natural habitats like forests to create farmlands, are killing these organisms, leaving soil vulnerable to erosion. Rising global temperatures which trigger droughts and wildfires are another factor. While more than 80 percent of farmland is used to raise and feed livestock globally, animal products provide only 18 percent of the calories we consume.

Today, over half of the people living on Earth cannot afford to eat a healthy diet; almost a billion people go to bed hungry each night, while billions live with the consequences of micronutrient deficiencies, which weaken immune systems and lead to preventable diseases. By 2050, if our population reaches 10 billion people, we will need to produce 60 percent more food to feed everyone. Food systems are already stressed under the pressure of changing weather patterns, extreme weather events, degradation of the soils and waterways and the disappearance of pollinators like bees, which play a crucial role in ecosystem health. While some claim that farmers produce enough to feed 10 billion people, we fail to manage our food efficiently; costs continue to rise, driving millions more into extreme poverty, hunger and malnutrition.

And we are not the only ones impacted. The Anthropocene is associated with an ongoing extinction event of species spanning numerous families of bacteria, fungi, plants and animals. Ecologically, humanity has consistently preyed on the adults of other apex predators, having a global effect on food networks. The globalisation of capitalism, coupled with the deregulation associated with neoliberalism, has accelerated the exploitation and destruction of the biosphere, resulting in the fastest mass-extinction of species in Earth's recent history. Human population growth, increasing consumption and meat production are the primary drivers of the mass-extinction, while deforestation, the destruction of wetlands, overfishing and ocean acidification are a few specific causes of global biodiversity loss.

In the US, twenty-two species of birds, fish, mussels and bats and one species of plant were declared extinct just in 2021; and in countries that don't collect such data or make it public, we are left to wonder about the extent of the damage. The vast majority of extinctions are undocumented; the current rate is estimated at 100 to 1,000 times higher than natural background extinction rates. And we still don't really know how many species we share the planet with; upper estimates indicate that between 10,000 and 100,000 species are becoming extinct globally each year. The most abrupt and devastating ecological effects of climate change can occur when environmental conditions become intolerable for several coexisting species simultaneously, causing chain-reaction die-offs.

In terms of what we do know: three-quarters of all living animal species could potentially vanish in the next couple of hundred years. The direct killing of megafauna for meat and body parts is the primary driver of their destruction, with 70 percent of the 362 megafauna species in decline as of 2019. Contemporary assessments have discovered that roughly 41 percent of amphibians, 25 percent of mammals, 21 percent of reptiles and 14 percent of birds are threatened with extinction, which could disrupt ecosystems on a global scale and eliminate hundreds of millions of years of phylogenetic diversity.

Mammals in particular have suffered such severe losses as the result of human activity that it could take several million years for them to recover. Relative newcomers to the planet and well represented in the fossil record, mammals emerged around 200 million years ago and are estimated to have had an average extinction rate of less than two species per million years. In the past 500 years, however, at least eighty of 5,570 known species of mammals have become extinct. A study in 2018 indicated that of all mammals, including humans and domesticated livestock, wild mammals account for only 4 percent of total mammal biomass on the planet. Various species are predicted to become extinct in the near future, among them the rhinoceros, non-human primates, pangolins and giraffes. Have we forgotten that we too are mammals?

Nowadays, over half of our population enjoys the luxury of the sum of human knowledge in our back pockets, and by that I mean our smartphones. So I am going to assume that we are already familiar with the range of ways in which we are accelerating towards our own demise. But let's summarise anyway.

The Anthropocene. Our population on planet Earth has doubled in the past fifty years, and the industry required to support our profit-driven, consumerist culture has destabilised our life-support system; we have plundered Earth's finite resources, disrupting habitats, eradicating species and polluting our own supply of water, soil and air at a rate at which the planet is no longer able to replenish. A mass-extinction event is currently underway, with up to a million plant and animal species predicted to disappear shortly. A particularly hard-line view may be that the Anthropocene Extinction is just a necessary by-product of human progress – but what is the point of such progress if we are no longer around?

And climate change is just the tip of the iceberg: we haven't even got to our being statistically overdue for a major asteroid impact, a geomagnetic field reversal, a solar superstorm as well as a supervolcanic eruption. And in fact the last two are not uncorrelated: data indicates that solar wind, via its disturbance of the Earth's magnetic field, has a strong influence on seismic and volcanic activity. Since 1755, sunspot recording indicates that the Sun's activity varies on cycles of approximately eleven years. The most recent, Solar Cycle 24, was the weakest in 100 years. Periods of activity produce more sunspots, as well as flares and coronal mass-ejections that trigger geomagnetic storms, auroras and potentially seismic and volcanic activity here on Earth. May 2024 has been the most active period of the current Solar Cycle 25 so far, and with the peak predicted for the years 2025 and 2026, it remains to be seen what the combined effect of current global warming and a strong solar cycle may be.

Whether natural or due to our activity or both, the disruption of our planetary environment, besides causing a mass-extinction event, also serves to increase poverty, inequality and conflict; not forgetting that our combined nuclear capabilities are enough to destroy humanity and all other terrestrial life fifty-five times over. It's worth pointing out that environmental disruption is also associated with an increased probability of epidemics; besides the obvious implications for our health, we have already witnessed the economic havoc wreaked on much of the world by the COVID-19 pandemic and associated shutdowns (which, interestingly, was the most profitable period for US-based companies since the Second World War). Capitalism, Karl Marx predicted just over 150 years ago, would inevitably concentrate wealth in the hands of a few while impoverishing everyone else; and in the aftermath of the pandemic shutdowns, and in the midst of the ongoing international and civil conflicts around the world, many over dwindling oil reserves, hundreds of millions more people are now living in extreme poverty than a few years ago.

Today, nearly a billion people live without power; almost 2 billion in inadequate shelter; more than 7 billion breathe polluted air; over 2 billion drink unsafe water; nearly a billion are undernourished; more than 2 billion don't have Internet access; and more than half of the world, that's 4 billion people, live without access to basic healthcare services. This is the most urgent challenge we have ever faced, because access to reliable power, adequate shelter, clean water, nutritious food, healthcare and connectivity are necessary for any human to acquire the skills with which to participate meaningfully in our society.

In its current form, our civilisation is not sustainable for much longer, and the state of Earth – our life-support system – reflects this. Our current system of managing our resources is no longer serving us: one study indicates that if we all lived like the so-called developed world, we would require several Earths to supply the natural resources. Neither does it faithfully represent our nature as living beings or as humans. Our society, our biosphere, our planet, like any complex system, is vulnerable to destruction, to the vacuum of space and the inevitable path through time into disorder that threatens not only humanity but all life forms which have evolved over the past 4 billion years. What can be built can far more easily be destroyed.

In short, the Anthropocene is characterised by a lack of respect for life, which is a result of the delusional belief that we humans are somehow separate from the natural world in which we live. And to think that ours is the only planet we know of where life exists! How can we remember who we really are and evolve a path for humanity whereby our society celebrates our curiosity and creativity, our sense of community and adventure, and the network of life that we depend on for our very existence?

Sometimes we need to zoom out to see the bigger picture. As a stern reminder, both our neighbours Mars and Venus once had bodies of water on their surfaces; we now observe average temperatures of less than negative 60 and almost 500 degrees Celsius respectively. This change in surface conditions is due to influences including alterations in atmospheric composition. But let's take a breather. On the bright side of the Anthropocene, we are learning something about terraforming. And while some of the most consequential changes to our world have been unintended, there are ways in which we can steer our course to be more purposeful – to heal rather than harm, in this world and potentially the next one too.

Terraforming

Terraforming is the hypothetical process of purposefully altering the atmosphere, temperature, surface topography or ecology of a celestial body to be habitable for terrestrial life including humans. There is a long history of discussion in science fiction, and more recently peer-reviewed research, on how we might modify the atmosphere of a planet, usually Mars, so that humans, animals and plants may live without protective gear on the surface. H. G. Wells may have made the first allusion to terraforming in his book *The War of the Worlds* in 1898: the Martians carried a quick-growing edible plant to Earth (whose growth had some unintended consequences). We've gleaned vast amounts of knowledge about our Solar System since then; so what are the prospects of transforming a celestial body in our neighbourhood into one we can walk around without sophisticated gear?

Given the surface conditions on some of the celestial bodies beyond Earth in our Solar System, a few global adjustments would need to be made before any terrestrial life, including plants, could be established there. For each world, the characteristics of its atmosphere, its distance from the Sun and period of rotation are primary factors contributing to temperatures there, while the strength of its magnetic field and the atmospheric composition determines the level of protection on the surface from cosmic and solar radiation. In situ resources include gases in the atmosphere, the surface materials as well as the incident sunlight. The basic requirements for the existence of terrestrial life include warmth, water and nutrition. Let's revisit some of the places we've become acquainted with in the search for life, but this time armed with hypothetical terraforming tools.

While Mercury has plenty of sunlight energy available, gravity almost the same as that of Mars and a core that produces a magnetic field (albeit just a percent of the strength of Earth's), its proximity to the Sun prevents the maintenance of an atmosphere. Even if a breathable atmosphere could be created, and positions near the poles with moderate temperatures found, the weak magnetic field means that high doses of sterilising radiation from the nearby Sun, as well as cosmic rays, will eventually exterminate life on the surface.

Venus has a weak effective magnetic field, not produced in its core but induced by the interaction of the solar wind with the planet's outer atmosphere and the charged particles there, as well as a thick atmosphere to protect it from radiation. However, with Venus' surface reaching temperatures of 500 degrees Celsius, we'd also need to think about modifying the 96 percent carbon dioxide atmosphere at a pressure of almost 100 times

Earth's at the surface. Carl Sagan proposed in 1961 that genetically modified microbes be introduced to absorb carbon dioxide and produce organic compounds; however, this process requires hydrogen, which is rare on Venus, mostly lost to space because of its high temperatures and weak magnetic field.

Establishing a human presence on the Moon is scheduled for this decade, relying primarily on the proximity of Earth but also to some extent on in situ resources, which include water ice in permanently shadowed craters near the poles. Without an atmosphere, temperature ranges are extreme; and while Earth's magnetic field to some extent shields the Moon from solar wind, radiation is more than 100 times that at Earth's surface and a couple of times higher than those experienced by astronauts aboard the ISS. With these radiation levels and the Moon's small mass, no matter how many resources we throw at it, any induced atmosphere will quickly disappear; so crews will likely be rotated frequently, perhaps on shorter cycles than the six months or more typical of stays on the ISS.

While living beneath the icy surface in the ocean of Europa or Enceladus would mean good radiation shielding, the cold temperatures, lack of atmosphere and intense radiation mean that terraforming their surfaces would be difficult. Ganymede is the largest moon in the System and has a strong magnetosphere. Also, radiation levels and gravity there are similar to those of our Moon, with the additional benefit of a thin oxygen atmosphere, as well as surface water ice as a potential source of more oxygen. However, to warm up the surface from current averages of around negative 150 degrees Celsius we'll need to think about how to introduce the huge amounts of greenhouse gases required to warm things up so far from the Sun, and how best to generate power there. Titan has a thick atmosphere, at around 50 percent Earth's pressure, and maintains an induced magnetosphere, like Venus. It's rich in resources, covered in liquid hydrocarbons such as methane and ethane, and potentially harbours a subsurface ocean of water and ammonia. Could we introduce enough oxygen from the water ice there to make the atmosphere breathable?

The problem with all of these Outer System moons is that they're years away with current propulsion systems, and we haven't yet mastered landing technology on their surfaces, never mind crews, who would then have to engage in the heavy industry required for terraforming, with little sunlight for power and communication delays of hours with Earth.

So, saving the best for last ... if we're going to try to terraform a celestial body, Mars is a good place to start. And sci-fi writers will agree; after all, we have surface knowledge acquired over decades, with our rovers giving us a detailed picture of the conditions and composition there. It's also just a few months' journey or ten light minutes away, and close enough to the Sun to receive about half the sunlight of Earth and maintain temperatures more moderate than any of the bodies considered so far. The major issues are the lack of a magnetic field, a thin, unbreathable atmosphere composed almost entirely of carbon dioxide, and radiation levels dozens of times those on Earth.

One way to terraform Mars could be by mimicking the only example of planetary-scale atmospheric oxygenation we know of; that induced by the proliferation of oxygen-producing microbes on the surface of Earth. For this we would need liquid water, which requires a higher temperature and pressure than currently exists on Mars. If we have learned anything from the Anthropocene, or from studying Venus for that matter, we will understand that one way to increase temperature, and simultaneously pressure, is via the introduction of greenhouse gases into the atmosphere. Can these be locally sourced?

Earth maintains the range of conditions of pressure and temperature within which the three phases of water, ice, water and steam, can exist; the triple point of water is at 0.01 degrees Celsius and just over 600 Pascals – interestingly, similar to the average pressure on Mars, equivalent to being some distance above the Earth's surface in the stratosphere at less than a percent of sea-level pressure. The stability of liquid water under Earth's surface conditions has significant implications for our planet's climate and the existence of life. Instead of focusing our terraforming endeavours on creating an Earth-like atmosphere with the rather anthropocentric aim of being able to breathe the air, let's relax the requirements – recall that Earth's atmosphere has only been breathable to mammals for less than half of its existence – and look at increasing the stability of liquid water. In fact, Mars is already on the edge of this range. Let's have a look not only at the phase transitions of water but also at carbon dioxide on the planet.

The fundamental factors underlying the chemistry of climate are pressure and temperature. Atmospheric pressure on the surface of Mars varies widely, from near-vacuum on the top of the largest volcano in the Solar System – the Olympus Mons, at altitudes of over 20 kilometres above the Martian plains – to around a percent of terrestrial sea-level pressure in low-lying areas like the impact crater Hellas Planitia. Pressure also varies seasonally with the fluctuation of carbon dioxide in the air; for the current average pressure on Mars the temperature range for liquid water is rather narrow, from zero to just over 2 degrees Celsius – meaning that liquid water, if ever present, is unstable and not there for long. If we double the pressure, to levels just more than in Hellas Planitia, this increases the window to between zero and nearly 4 degrees Celsius. Could we nudge Mars towards a situation where this window is wider? Increasing the temperature or the pressure even by a few percent would suffice.

If we could find a local resource on Mars that, when added to the atmosphere, would react with the carbon dioxide there to form a powerful greenhouse gas stable under UV radiation, thereby reducing the amount of ambient carbon dioxide that freezes in winter, we could widen the temperature range at which water can exist. And if there are pools of water maintained anywhere, carefully selected or genetically modified microbial populations could be added to these environments, emitting more gases that will further increase the temperature and pressure. Because of the tilt of the planet's rotational axis, summer in the south is winter in the north, and vice versa, just like Earth. And, just like Earth, Milanković cycles could inform us of the most efficient timing to release such a gas.

Then there's still the issue of the lack of magnetic field to protect the thicker atmosphere we may try to create, as well as to protect any life on the surface from harmful radiation levels. Both Venus and Mars, while lacking intrinsic magnetic fields, are surrounded by a plasma environment which plays a role in protection from the solar wind, as well as the loss of atmospheric content into space. Our knowledge of the rate and mechanism of ongoing atmospheric loss on Mars has had a significant contribution from the Indian Mars Orbiter Mission of 2014. India was the first country to enter Mars orbit at the first attempt (followed by the UAE, which also entered orbit at the first attempt in 2021). Interestingly, the cost of the Indian mission was about 75 million US dollars; at around the same time, a Hollywood movie about space came out called *Gravity* which had a budget of 25 million more. But back to the Martian atmosphere, or rather lack thereof. Venus' induced magnetic field plays an important role in maintaining its thick atmosphere, deflecting particles from solar winds and protecting the atmosphere from being stripped away. Mars also has a weak induced field, but could we boost it? Considered more fictionally than scientifically in films like *The Core*, one option that has been suggested is to restart Mars' internal dynamo. Besides the bombastic amount of energy required, though, there are no guarantees that melting Mars' core will kick-start an internally generated magnetic field; assuming this was the origin of Mars' previously existing magnetic field, we still don't know what halted it.

Let's for a moment assume that there is some feasible transformation from the current conditions on Mars – perhaps by the addition of some hypothetical locally sourced substance that, in modest quantities and with well-timed release, plays the role of a powerful greenhouse gas in the upper atmosphere – into a place where pockets of liquid water can be maintained on the surface. Let's not stop there. Perhaps this new atmospheric composition also boosts the conductivity of the ionosphere that interacts with the solar wind to increase the strength of Mars' induced magnetic field; and under this newly robust protection, microbes are able to survive in the open, emitting gases potentially including oxygen from the surface pools of water that are now stable for much of the year in low-lying regions. I don't think I need to remind you that this is highly speculative. Then, microbial populations could potentially flourish on the surface of Mars and begin to terraform the planet.

But is this not reminiscent of the same hubris with which we have eroded the functionality of our life-support system here on Earth? What of the unknown feedback loops, in particular for the as yet unknown Martian life potentially eking out an existence beneath the surface? Creating a living world on the surface of Mars is a grand vision, but for now we – just like possible indigenous organisms – are probably limited to living underground. And probably for the best.

Look at us theorising about making other planets habitable for terrestrial life while we continue to disrupt the balance of the biosphere and destroy our fellow creatures who inhabit the 4-billion-year living library here on Earth! Who can deny that humans are crazy dreamers? What defines us are our aspirations, our ability to imagine worlds beyond the reality we are experiencing, which constantly drive our curiosity, our resilience and our need to explore. We can prevail in spite of ourselves. We are built for it.

Staying on one planet is a sure and inevitable path to extinction. Even if we by some miracle make it through the next billion years of natural disasters, climate change, pandemics and war, the Sun will by then be 10 percent brighter, and a runaway greenhouse effect will scorch Earth and all remaining terrestrial life. All known living species, and all evidence thereof, will be gone forever. With the weight of 4 billion years of evolution and complexification on our shoulders, is it not our responsibility to expand the precious light of life to places of safety before this occurs?

As a child, I read a lot and longed to live in another era, when less was done, more was at stake and important pathways for humanity were yet to be trodden. I imagined myself as an Egyptian queen or a Viking warrior, anything but me, at the tail end of the twentieth century, stuck in 'the Pit' at the bottom of Africa.

However, I am now convinced that the time in which we are living is the most important of all: 200,000 years of exploration, innovation and the shaping of our environment on a global scale have resulted in the destabilisation of our life-support system here on Earth, the only one we have ever known. We are approaching a critical juncture for terrestrial life, and in particular for humanity. In the past we see rich complexity emerging from such planetary-scale upheavals. But will this complexity include us? Our ancestors had to adapt to scenarios beyond their control; this time we have the added dimension of the impact our own activities are having on our global climate. We'll need to harness all our history of collective knowledge to get through this one. But if we do, a whole new transformed world awaits.

We have seen how exploration has shaped our society, how our fundamental characteristics of curiosity and creativity have enabled our expansion across the surface of our planet, while at the same time our innovative tools and capability to shape our environment, coupled with our exponentially growing population, have driven the destruction of the only home we have. How can we reconcile the current state of affairs on Earth with the belief that we can do better? Where to from here?

PART III

WHERE ARE WE GOING?

I was there in Mexico in 2016, when founder and CEO Elon Musk unveiled SpaceX's Mars programme. It felt like queuing for a rock concert or soccer game, except that people were drinking coffee and it was before noon. When the doors opened, they started shouting and running to get seats. I stood on the side, away from the stampede, to be near the front. And then Musk walked on stage and thousands of space fanatics cheered.

I breathed in the moment and thought of Ragnar Lothbrok or Christopher Columbus pitching plans to sail west in search of new territory to their respective monarchs. But here was a man with the ability to implement his plan; a multi-billionaire with a business model and technical designs for the most ambitious project ever proposed in the 4-billion-year history of life on Earth: making humanity a multiplanetary species.

As Musk has put it, Mars is a 'fixer-upper of a planet'; it's also a natural stepping-stone into space for humanity from Earth. A few months at sea didn't put our ancestors off, and as it is the less hostile of two neighbouring planets, a crewed mission to Mars is an entry-level requirement towards becoming a space-faring civilisation. As the first private company to deliver cargo and crew to the ISS, as well as achieving reusability of launch systems through routine re-landing of rocket boosters both on land and at sea, SpaceX is on a steep curve of success to achieve Musk's goal of establishing a self-sustaining society of around a million people within 100 years on next-door-neighbour Mars. (SpaceX may not be alone: while China is achieving milestone after milestone in its ambitious space exploration programme, the UAE has also set its sights on a century plan for a city on Mars, announced in 2017. And if anyone can build large-scale infrastructure in a desert from scratch, China and the UAE can.)

Assuming that the development of SpaceX's Starship continues successfully, and that NASA's Artemis programme provides the opportunity for SpaceX to demonstrate crewed landing capabilities off-world, then a few 100-tonne cargo deliveries are all that stand between us and the first crews' arrival on Mars. But what then?

Unlike our ancestors, who set off towards the horizon without knowing what lay in store, we have detailed knowledge of the surface of Mars owing to decades of data acquired from Earth-based and orbital observations, as well as landers and rovers with direct experience of the ground conditions. While we have not yet brought back any samples from Mars, hundreds of meteorites retrieved on Earth have been identified as Martian, further contributing to our knowledge of the planet. Nowadays, large data sets – including high-resolution video footage taken by rovers on the ground as well as, more recently, a drone flying above the surface – are expanding our imaginations all the way to the surface of Mars.

One night, on a ten-day silent Vipassana course some years ago in the mountains outside Cape Town entailing eleven hours of meditation a day and not much else, I woke up utterly convinced I was on Mars. I could hear the hum of the life-support system and feel the rest of the crew sleeping close by. I had a clear sense of what it felt like to be so far from home, on the surface of another planet, waiting for day to break – until I eventually found the light switch revealing my simple room at the retreat.

The four remaining people alive today who know what it's like to be on the surface of the Moon will not be around for very much longer, being at the time of writing in their late eighties and nineties. With high-resolution visuals streaming into our consciousness from rovers on the surface of Mars, can we imagine what it would be like on the surface of this world within our reach? Let's say I wake up on the surface of Mars as a member of a community living there; what would a 'day in the life' look like? For now, it may seem a bit like science fiction. But for those ready to expand beyond this world and into the next, let me take you on a tour of a vision that could become reality in as little as the next decade: welcome to a day in the life on Mars.

7

OFF-WORLD

A day in the life on Mars

I wake up early, the rest of the team still asleep, to prepare for the guests arriving from Earth today. I walk quietly through the shared sleeping area, through an airlock and into a corridor, touching the uneven walls of the faintly lit tunnel as I go, feeling the smooth sealant layer with which we have coated the walls of the subsurface lava tube. By adding urea, yes from our urine, to experimental regolith composites made from Martian sand, we improved the strength of this epoxy to perform well as an airtight coating, preventing gas leakage in or out of the porous surface of the subterranean cavern walls. Sleeping and doing some of our work underground provides protection from the relentless cosmic and solar radiation – the thin atmosphere doesn't afford much protection – on the surface of this home away from home. I pass through another airlock and then climb, rather easily in 0.4g (which is gravity just more than a third of Earth's) up to the surface facility.

All is still, besides the faint buzz of the air conditioning systems. I gaze through the large window on the viewing deck at the yet-dark night sky; the same stars are visible as from Earth. However, two oddly shaped and less familiar small moons are also visible to the naked eye, Phobos and Deimos. There is a blue dot in the vast night sky too: Earth. Although the new world just outside is so close, at night temperatures can plummet below negative 90 degrees Celsius – the coldest temperature ever measured on Earth – and going outside without the correct gear would result in hypothermia, unconsciousness and then organ failure within minutes. At pressures 100 times less than sea-level pressure on Earth, exposure will immediately affect the eardrums and cause exposed moisture to evaporate, impacting vision and reducing the ability of the lungs to absorb oxygen, which is only present in minuscule amounts anyway. To prevent rapid death, we can only experience Mars through a protective suit. While on Earth we coexist with abundant liquid water and plant life that prevails in the ambient conditions, and are able to sense the molecular components of the environment as we breathe in the air, we'll never 'smell' Mars, not in the current state of the atmosphere anyway. And certainly not anytime soon. Unless this week's meetings go particularly well, that is …

We haven't yet celebrated a coffee bean harvest; I add leaves from our camellia sinensis *plants to the hot water in my cup and sip on green tea. The steam mingles with the scent of the freshly printed composite wall panels of the inner viewing deck and the warmth of the ventilation systems, keeping the deadly outside environment at bay. Gradually, a glow appears in the east. The solar arrays glint in the first light. Sunrise has a blue hue because of light's interaction with very fine Martian dust particles. The horizon looks close; Mars is just over half as wide as Earth. The sand is a reddish colour because of the iron oxide, or rust, content, which creates a pinkish sky as the Sun begins to rise higher, appearing smaller, being around 50 percent further away than from Earth. Light winds stir up the fine dust, which falls back down to the ground at an unfamiliar rate in the low pressure and reduced gravity environment.*

I am grateful for the tea, all locally sourced, besides the seeds that is. The surface sand on Mars contains traces of water ice, with larger deposits deeper under the ground; there may even be liquid water maintained by geothermal heating and pressure beneath the surface, particularly further south in the polar region, which remains to be investigated. Our landing site was selected both owing to the estimated availability of near- and subsurface ice as well as the distance from potential liquid water deposits. While all known life, including our humble multi-species community, requires the precious H_2O molecule for survival, we also want to prevent contamination by our activity of such potential underground lakes which may host Martian life. Some of our researchers were part of the team that detected microbial life living off methane and ammonia in lakes in the dark under a kilometre of ice in Antarctica.

Meanwhile, the gentle morning Sun has begun to illuminate the sloping set of craters to the west (to whose past volcanic activity we owe our subsurface shelter), the canyon gouged out to the east, and the various rock features and erosion patterns, all of which indicate the movement of large bodies of water in this region between the southern highlands and the low plains to the north. While Mars once had oceans, current average surface conditions do not support liquid water, which sublimates directly from ice into gas. It's now the beginning of summer, which lasts around six Earth months, and we expect temperatures to reach perhaps 20 degrees Celsius during the day. As the surface begins to warm up in the morning sun, the solar-powered rovers awaken and slowly commence their dredging in the sand fields. The sand is scooped up and dropped into subsurface containers, where it is heated to collect the emitted water vapour; a tonne of sand can produce a few dozen litres of water, which is then purified. The remaining sand is used for materials production, including resins for sealed expansions of our subsurface habitat as well as construction materials for the surface facilities, so far consisting of an interconnected network of inflated domes brought from Earth as well as more permanent structures built with Martian sand composites around our initial home in the spacecraft lander.

Water is a fundamental resource that besides enabling life in the form of, for example, tea or irrigation also provides the ingredients for breathable air and fuel. Even *if the Martian atmosphere contained enough oxygen, it would be unbreathable due to the low pressure. As it is, the Martian atmosphere is around 96 percent carbon dioxide, containing also small amounts of argon and molecular nitrogen, and less than a percent of oxygen. I finish my tea and take a deep breath. I am looking forward to showing the new arrivals what we have managed to achieve here in just under a year. Time to check on the life-support systems.*

I wake up a series of large touch screens on the inside wall of the viewing room. The most fundamental process keeping our community alive on Mars is power genera-tion. Energy availability enables all other functionalities of the settlement. If power fails, within minutes so does life; hence redundancy of power-generation systems is critical.

The Sun, a most steadfast source of energy in the Solar System, also reliably disappears for around half of the day–night cycle, which coincidently on Mars is just a bit longer than the twenty-four-hour cycle on Earth. And so at night we mimic the Sun, producing power from the energy stored within the nuclei of atoms. While we're excited to follow the rapid progress of fusion power-generation capabilities on Earth, the mass and size of the technology, for now, remains prohibitively large for transportation and use off-world. Our modular thorium fission reactors, on the other hand, run reliably for several years on less than 100 kilograms of fuel brought from Earth, with the capacity to provide for all of our power needs, for now anyway.

Molten salt thorium reactors are not prone to explosion or meltdown; however, a widening in the lava tube of a few hundred metres and several reinforced airlocks down from the greenhouse cavern felt like an appropriate distance for the reactor room. In addition to enabling efficient power production, the high operating temperatures of our reactors also provide heat for processes like materials synthesis and warming the greenhouse. Our long-term goal is to extract thorium from nearby surface deposits identified by satellite analysis with an abundance of perhaps as much as a part per million. We also have a range of nuclear batteries of various capacities and levels of portability; with no moving parts and with appropriate shielding, these are reliable sources of heat and electricity that were immediately functional on arrival, and are also suitable for transportation to power more distant outposts as we engage in remote exploration away from the base.

The readings on the reactors are all nominal. I swipe to a different dashboard; an overview of power allocation. The Sun is still low in the sky, but even by midday the solar irradiance on the surface of Mars is at best around half of that on Earth. In any case our strategy has been to use as much solar power as possible during the day. We are evaluating the lifetime of the various panel designs in our arrays in the harsh surface conditions, in particular the radiation, towards our objective to manufacture panel components locally from in situ resources. The readings indicate that as some systems switch over to solar as the Sun rises in the sky, various fuel-production processes have been initiated to store the excess electricity produced by the nuclear reactors.

Not only for energy storage but also as part of our resource cycling (no such thing as waste here!), we use a range of fuel cells because of the by-products. For example, a regenerative fuel cell system can store energy via the electrolysis of water to produce oxygen and hydrogen, with the energy-generation phase producing water via a reaction between these gases. And so we can produce either hydrogen and oxygen, or water and electricity, depending on our needs. Other cells utilise a range of organic materials for fuel, including wastewater. And the output, besides electricity? Cleaner wastewater. Which we process further in varying degrees depending on the intended use. Another by-product is nitrogen.

While water is extracted from the surface using rovers and heating procedures, carbon dioxide and small amounts of other gases are pumped directly from the Martian atmosphere. With oxygen produced by electrolysis from either of these molecules, creating pressurised breathable air additionally requires an inert gas like nitrogen, which makes up just a few percent of the Martian atmosphere. I tap on the air icon. We decided to maintain a mile-high atmosphere, similar to cities like Johannesburg or Denver, with Earth-like composition to minimise fire hazard as well as unforeseen physiological *responses to long-term exposure to novel mixtures; all is normal with readings around the base indicating 78 percent nitrogen, 21 percent oxygen, under a percent of argon and carbon dioxide maintained between 0.04 and 0.1 percent.*

Through the controlled use of specific microbes we can retrieve both pure water as well as nitrogen gas and nitrogen-based compounds used for example for fertiliser from the digestion of organic waste. We are currently experimenting with various microbial colonies brought from Earth to optimise the efficiency of bioelectricity generation via microbial metabolism, as well as to direct the production of specific by-products we require. I go down a level on the power allocation dashboard; all of the water systems are functioning normally. With increased power assigned to water production over the past few weeks, we have expanded capacity for the new arrivals, who would have been wet wiping for the duration of the 100-day flight over here and will probably enjoy a shower. We decided to locate all activities utilising water, like washing, in sections of the greenhouse, to keep the humidity where it's needed most. And because taking a shower in the bamboo grove is something we all look forward to.

Critical for the transformation of resources is the oldest living process on Earth, photosynthesis. In the greenhouse, or 'the garden' as we call it, we grow food for our community in agricultural units with specialised environments optimised for each group of species living there. The majority of our daily dietary requirements from carbohydrates to proteins to essential nutrients come from the garden, a lively ecosystem home to things like: berries, leafy greens, cassava, mushrooms, insects and fish (conveniently transportable from Earth in egg form), bees that were brought live in a hive after much debate around flight safety, algae and cyanobacteria including spirulina grown in photobioreactors, also a range of medicinal plants, fungi and bacteria as specific sources of biomolecules for healthcare applications among others. We have over a thousand species of life in our ecosystem, undocumented microbial hitchhikers aside, and we monitor the health of each as well as the entire ecosystem with great care.

In the garden, we collect data on health parameters for each species, and are able to adjust the environment towards optimal well-being. We can control factors including the strength of the artificial magnetic field (which we found to be critical for plant growth and stress tolerance as predicted by quantum biology research), the spectrum and intensity of light emitted by our LED arrays, the composition, pressure, temperature and movement of the air, the nutrient content of the water in the irrigation system, the proximity of species to each other (some enjoy cohabiting), as well as, critically, the timing for each of these resource deliveries, on the Martian clock with a day of twenty-four hours and thirty-nine minutes. We've found some interesting results.

With a range of sensors at our disposal, some of our researchers have been analysing the garden data sets, in particular the electromagnetic emissions. What we have found is that in the near ultraviolet through to high frequency visible range, there is a set of ambient characteristic frequency patterns, detectable in all parts of the garden, that are correlated with what appear to be states of optimal health of the ecosystem; salutogenesis as we call it. With almost a year of detailed data on the garden ecosystem and sufficient computing power to run machine learning algorithms, we are now able to remedy the majority of deviations from salutogenesis automatically by adjustments of resource inputs; but the real surprise was that humans gathering in the garden to share food or discuss community issues can also impact this spectrum, in some cases resulting in a shift towards salutogenesis even without any adjustments to, for example, the lighting or irrigation systems. Life likes life! We didn't need any further excuse to hold most of our mealtimes and meetings in the greenhouse area.

We also have a systems health measure for our entire settlement; and now is a good time to introduce our resource-management system. Our basic resource infrastructure – I swipe back to the power generation dashboard – consists firstly of energy systems: thorium nuclear reactors; nuclear batteries; solar arrays and fuel cells. Since energy is required to generate water, at the next level, we have water systems; from our rovers collecting ice-containing sand to our wastewater-processing *facilities, monitoring water production and utilisation is critical beyond just applications of the H_2O molecule, for keeping track of oxygen and hydrogen availability. Requiring both energy and water, at the next level is the garden, and in the final tier, relying on all the hierarchy of resource-production systems, is our human team, with real-time physiological health indicators based on the outputs of various wearables constantly monitored (behind some privacy settings) for homeostasis. While these tiers provide one way to look at it, the underlying resource-management system is complex; each unit of equipment has degrees of relation to each other unit, forming a life-support network that realises an efficient flow between energy and matter to keep our precious biosphere thriving. Let's look at an example.*

The visitors arriving today will need to refuel before returning to Earth. Only while all life-support systems were functioning normally, we began allocating additional power to stockpile rocket fuel, as well as water, some months ago. One kind of fuel can be produced by pumping carbon dioxide from the atmosphere, introducing it into one of our photobioreactors, where it's converted by cyanobacteria into sugars, and feeding these sugars to our lively colony of E. coli that then produce a kind of fuel called 2,3-butanediol; a couple of rocket designs using this fuel are currently being developed on Earth. But our visitors are arriving in a methane-burning vehicle. We can produce methane, CH_4, by anaerobic digestion, a process through which complex microbial *communities break down organic matter, including our food waste and sewage, in the absence of oxygen in a sealed reactor. The methanogenic bacteria digest the input and produce biogas, which is almost two-thirds methane, also carbon dioxide and trace organic gases, and a nutrient-rich leftover material, or digestate, for which we have various uses in the garden. Another method of methane production is via the Sabatier process, used for example in space stations to scrub the carbon dioxide from the air astronauts breathe. The chemical process entails a reaction of hydrogen obtained via the electrolysis of water with carbon dioxide at a high temperature to produce methane and water. Our optimisation experiments on microbial community composition and the use of catalysts such as graphene aside for now, each of these methane-production processes has an energy price tag. So which one do we use? Let's take a look at the principles on which our resource-management system is based.*

Our data-driven, resource-based economic system is founded on some basic priorities: the continuity of life; efficient use of energy; and transparent resource distribution. Yes, that's pretty much it. We call our resource-management system, basically a primitive economic system without the money, DataS. The name originated from some kind of joke around the pronunciation and plural of 'data' and a Star Trek reference, and stuck. At its heart, DataS is dedicated to keeping us alive; for this, we also require the good health of the species we have brought with us. All living beings need energy and resources, so we aim to use these as efficiently and equitably as possible to optimise the health and longevity of our biosphere. Defining equity turned out to be a heated philosophical debate with no clear resolution in sight, so we decided transparency of resource distribution will suffice for now. Since all resource production and processing requires energy, the simple explanation of DataS is that it issues levels of priority for the various resource-transformation processes necessary for life based on the optimisation of energy efficiency.

On a basic level, DataS calculates and recommends the more energy-efficient of the two techniques of methane production; each process has an energy price tag considering the volume of methane required in a specific timeframe for refuelling. Additionally, the inputs and by-products of each process are also taken into account, for example we may have a surplus or lack of waste to add to the digester, or we may want to remove carbon dioxide from our air via the Sabatier process. The energy price tags of performing these required resource transformations is factored into the recommendation.

Energy availability provides an indicator of the fundamental health of the system, since energy is required for any transformation of resources from one form to another. In a way, energy is currency here. By monitoring power production, DataS provides continuous data on which subsequent decisions on power allocation for all other operations are made. The highest priority is maintaining the nominal function of the life-support systems required to keep our team and the organisms making up our biosphere alive. When we have surplus energy, we can allocate this to secondary processes like research, exploration or entertainment. And not always in that order, judging by the popularity of our first palatable batch of beer. While we use machine learning algorithms on all of the data generated from sensors that continuously monitor every aspect of the energy and resources being produced and utilised in our community, we have not outsourced decision-making to 'AI'; all decisions are agreed upon by us, mostly in the garden. So far, while we can all fit round a (rather large) table, we have been able to reach mostly unanimous decisions when DataS provides comparable priority for two different options.

Most kinds of equipment we can't yet produce locally; for surface suits and computer components, rovers and nuclear reactors and much more, we rely on resupply from Earth. In the longer term, we aim to locally manufacture the infrastructure to produce power and cycle resources as outlined above. Local and in particular additive-manufacturing is essential for emergency maintenance and repair of life-support infrastructure according to design data sent, if necessary, from Earth at light speed, and eventually we hope to produce things like 3D printers themselves, using recycled as well as in situ resources.

Martian innovation is already underway. For

example, graphene, a single-atom-thick sheet of carbon, conducts electricity better than copper; is stronger and lighter than steel; can replace indium in touch screens and a range of rare metals in solar cells; provides radiation shielding; and has a plethora of other applications in various industries. A few years ago a method to produce graphene by combining a simple hydrocarbon gas, oxygen and a spark plug inside a sealed chamber was discovered by researchers on Earth. Over there, the discovery was published as a research article and the experiment was repeated by a few other groups; even the discovery that graphene can be produced from the same microplastics that pollute Earth's oceans continues to gain traction there. On Mars, we are already using locally produced graphene by such simple methods in various electronic contexts, but perhaps the most fruitful area has been as a catalyst, in both our microbial fuel cells and methane digesters, and we are also experimenting with graphene as a component in solar cells as we work towards their local manufacturing.

I bring up the tab on exploration indicators. DataS recommends that we reduce our daily energy consumption at the base by a few percent before diverting resources for an expedition further south to scout for resources like thorium and subsurface ice, and also to investigate the dark, narrow streaks called recurring slope lineae flowing downhill on nearby crater slopes that orbital missions have identified, and may be caused by the flow of subsurface liquid water. And there's nothing like the presence of liquid water to get our biologists excited! While searching for life or direct evidence thereof here beyond Earth is the most important mission that we or arguably any human has ever embarked on, we have spent a lot of time establishing a stable biosphere to support this activity first. It looks like we will be ready for our first expedition in a few months. The anticipation is tangible.

Time to go to breakfast. I head back underground and down the passageway in the opposite direction to the sleeping area. I close my eyes as I open the airlock to the garden; it smells like Earth. The sounds of people talking and laughing and serving food and drinks permeates the humid air heavy with the scents of the multitude of different plants growing there, alongside the microbes, insects and worms participating in the lively ecosystem. After I eat some berries and herb salad from the garden along with some kind of egg derivative from Earth, DataS makes an announcement on the habitat-wide audio system – also conveniently repurposable for karaoke sessions – that the visitors already in Mars orbit will be engaging landing protocols and touching down shortly a few kilometres from our settlement. We clean up and move excitedly to the surface facilities, where the welcoming crew suits up and readies the transport rovers to meet the lander. The guests' vehicle appears as a dot in the sky, growing slowly larger, before touching down in a cloud of dust. With coast speeds of over 100,000 kilometres per hour, the guests have been travelling for

just over 100 days. We await confirmation of a success-
ful landing before starting the rovers and heading off
towards the landing site. I touch the rock in my pocket, a
quasicrystal from a deposit found in one of the caverns,
with the unexpected properties that might just make ter-
raforming Mars feasible …

For those waiting with bated breath for descriptions here
of space elevators or quantum AI or genetically engi-
neered Martian humans: sorry to disappoint. Virtually
all the technology we require to inhabit Mars already
exists, and this imagining isn't as far off as you might
think – sometime in the 2030s, perhaps! Besides the travel
distance, living in the ISS or even on the Moon is actually
more challenging from an engineering perspective than
setting up camp on a rocky planet with significant gravity and an atmosphere. Whether it's a novel mineral or
indeed some microbial community found in a niche environment like a lava tunnel on Mars, there is no doubt that
a human population living on the surface of the planet
next door will make new discoveries, perhaps even far
beyond what we can imagine. It will, after all, be the first
time terrestrial life is exploring the completely new environment of another planet, and will certainly be a unique
set of humans who travel there to build a new world.

Whether or not we endeavour to shape the environment on Mars into one where liquid water can prevail at
the surface, or even one where we can breathe the air, if
we are to live there we are going to need resources. And if

we are to aspire to increasing independence from Earth – a wise move with resupply missions taking months, and even light taking minutes to traverse the distance between the worlds – we will need a lot of resources. Delivering megatons of materials and equipment from Earth is not sustainable, likely to be supported, or even feasible. Perhaps, conversely, Earth may be interested in the resources accessible from Mars.

Besides Earth, Phobos and Deimos, there are two other new objects visible to the naked eye in the Martian night sky: Ceres and Vesta, the two biggest objects in the Asteroid Belt. And while the infrastructure required to perform Earth-style subsurface mining may be a long way off for early Martian communities, those of us living there will be space-faring natives and, as humans, surely compelled to explore the surroundings of our new home.

The Belt and beyond

There are correlations between locations where significant amounts of heavy metals and rare minerals have been extracted on Earth and meteorite impact sites. When a large body, typically an asteroid fragment or a comet, crashes into Earth, it deposits its contents near the surface. Additionally, the immense amounts of energy transferred to the crust of the Earth during a collision can produce minerals, trigger the flow of metal-containing lava from beneath the surface, or redistribute or concentrate existing metals in a localised region.

For example, around 2 billion years ago a meteorite 20 kilometres wide smashed into an area just south of what is now Johannesburg in South Africa. The 300-kilometre-wide impact site, the Vredefort Dome, is the largest on Earth, and is co-located with the Witwatersrand region, where the largest deposit of gold and the largest rough diamond ever found on the planet were extracted. Today, this region is home to all four of the deepest mines on Earth, with operating depths of up to 4 kilometres as we exhaust the gold conveniently concentrated near the surface. Dozens of other impact sites around the planet have also been identified as localised sources of resources including minerals, metals and construction materials. Instead of digging up the surface of our new home away from home, what if we went directly to the source? Just beyond Mars lies the Asteroid Belt – home to an abundance of the 'scarce' resources required for this world and the next.

Too small to be planets, too big to be ignored, asteroids have been the cause of the extinction of many a species here on Earth due to global environmental changes that result when a big enough impact with the surface of our planet takes place. The characterisation of asteroids is therefore a critical first step towards mitigating extinction-level impacts; there are many near-Earth asteroids traversing the Solar System that are as yet uncatalogued.

Ranging in size from nearly 1,000 kilometres across, like Ceres, to dust particles, asteroids are by and large found in between the orbits of Mars and Jupiter, though some deviate with more eccentric orbits while a few pass close to the Earth, or impact the atmosphere as meteors. Besides mitigating impact hazards, compelling reasons to understand asteroids include enhancing scientific knowledge – understanding the origins of our Solar System and the formation of the celestial bodies found here – as well as, critically for our expansion beyond Earth, utilising their resources. The extraction of extra-terrestrial resources from space rocks can equip both the establishment of off-world communities and our continued exploration of the Solar System.

Most of the contained metals within large bodies like moons and planets are typically found towards the core, having gravitated there over time due to their relative density; rendering the information as well as the resources contained there largely inaccessible. On the other hand, some of the bodies in the Asteroid Belt are themselves remnant metallic cores or fragments of early-stage planets called protoplanets, formed when smaller rocks collided and melted under the heat generated. Some such protoplanets were then stripped of their surface crust and mantles by subsequent massive collisions typical of the early System. This is one interpretation of how the largely metallic so-called M-type asteroids, composing up to 10 percent of the Belt, formed. Most asteroids are classified as carbonaceous chondrites, constituting around three-quarters of the Belt and containing things like water, organic compounds, silicates, oxides and sulphides. The larger of these may have undifferentiated and solid composition, while most of the smaller asteroids are thought to be collections of rubble held together loosely by gravity; fragments of the original planetesimals. Within M-types, ferrous metals like iron are most abundant; but some metallic asteroids are estimated to contain, for example, hundreds of parts per million of platinum, orders of magnitude more than typical concentrations in the Earth's crust.

While our current activities beyond Earth, for example in Earth orbit and for planned lunar bases, rely heavily on resupply from Earth, to be self-sufficient in the longer term off-world communities will require significant amounts of resources to be sourced locally. The beauty of exploration is that once we are able to leave the finite environment on Earth, resources are more abundant than we think. And to extract these resources, we don't need to disrupt the only thriving biosphere that we know of: the living ecosystem unique to Earth. The answer to our future resource requirements, whichever planet we are on, may lie in the region just beyond Mars.

Only a few generations ago, less than a dozen materials were in wide use, among them wood, brick, iron, copper, gold, silver and a few plastics. By contrast, a single modern computer chip uses more than sixty different elements of varying scarcity. Earth is largely a closed system, losing just heat, hydrogen and helium to its surroundings. Bar very few extremely rare metals, the notion that we are 'running out' of certain resources is true only in the economic sense that extracting, recycling or transporting these resources is not cost-efficient.

Increased societal demand, coupled with our failure to implement efficient waste-management strategies, may result in it becoming profitable to return certain off-world resources to Earth. While the Moon and Mars have compositions not dissimilar to Earth, Asteroid Psyche, for example, is a 250-kilometre-wide lump of rock estimated to contain rare metals and minerals worth 100,000 quadrillion US dollars. This is more than the value of Earth's entire economy – if the current economy is even compatible with such an abundance of 'scarce' resources.

The high-level requirements to prepare a typical space mission are comparable with those to establish a terrestrial mine: large-scale infrastructure; teams of experts working for typically over a decade; costs running into billions of US dollars. A feasibility study in 2012 commissioned by the Keck Institute for Space Studies concluded that to retrieve a near-Earth asteroid of around 7 metres and return it to Earth orbit would cost 2.6 billion US dollars and take up to ten years. Meanwhile, after over a decade of planning and construction, Cobre Panamá, a 10-billion-US-dollar copper mine in the rainforest in Panama, is now idle due to mostly political reasons.

Which leads us to another point: what's left of potential locations for mining are mostly already home either to humans or to increasingly shrinking natural ecosystems. The total mass of the collection of rocks between Mars and Jupiter is estimated to be around 4 percent of the mass of the Moon; however, this region likely contains more metals and minerals than what's left of Earth's surface reserves, because among these rocks are fragmented planetesimal cores, containing the same stuff we've been digging deep into Earth's crust to access.

Thus far, government agencies have launched a number of missions to asteroids and comets: a US flyby mission collected particle samples from a comet's coma in 2004; a European mission was first to land on a comet in 2016; Japan has twice retrieved asteroid samples, in 2010 and 2020; and the US has recently returned asteroid samples to Earth, as well as testing asteroid deflection technology in the DART mission, with the unintended consequence of causing a meteor shower on Mars and potentially also on Earth in the coming years. Commercially, private space mining companies including Planetary Resources and Deep Space Industries were established a couple of decades ago, with the aim of extracting resources from asteroids. The passing of legislation in the US and then in Luxembourg allowing companies ownership of what they extract from celestial bodies did not, however, enable these two companies to succeed. Perhaps this was because the leap of imagination required from mining the surface of the Earth to drilling, blasting, cutting and crushing in the vacuum of space hundreds of millions of kilometres away was too much for investors to handle.

More successful ventures may be those that have application somewhere closer to home, like Earth orbit. Furthermore, we may need to rethink traditional mining methods. A candidate concept for asteroid resource extraction involving the encapsulation of the entire target goes by the acronym SHEPHERD (Secure Handling by Encapsulation of a Planetesimal Heading to Earth-moon Retrograde-orbit Delivery); scientist Bruce Damer is among its creators. The concept involves sealing a target, anything from a defunct satellite to a planetesimal, within a gas-filled enclosure, detumbling (these rocks are typically spinning at potentially high rates), and then redirecting it to the required orbit using gas flow.

Owing to the minimal interaction of the method, sensitive items like satellites could be targeted: data stores retrieved; refuelling or maintenance performed; or repurposing or recycling of defunct components achieved. Further away, many asteroids are aptly described as weakly consolidated rubble piles; their encapsulation for resource extraction would provide protection for spacecraft hardware from loose debris and dust disturbed on the target's surface, and enable the acquisition of volatiles from icy objects through heating and the processing of materials, for example via electroforming, on the way to the desired destination.

While bringing resources back to Earth could contribute to the cost-effectiveness of initial missions, the real potential of asteroid-mining is in realising humanity's ambition to explore and settle space. Launch from Earth is getting cheaper, thanks to companies like SpaceX, but the cost of getting large amounts of resources off of Earth's surface remains prohibitive as well as illogical. Finding and extracting useful materials in space could turn asteroids into fuel stops. The most important resource is water, which aside from its uses in life support and food cultivation also can be broken up into its constituent parts – hydrogen and oxygen – to create rocket fuel and breathable air. Discussions around SHEPHERD creating fuel depots for SpaceX's Starship are already underway. Other resources like organic compounds, minerals and metals extracted in space will not only extend the range of crewed space exploration; in combination with technologies such as 3D printing, resources extracted from asteroids can be used to create tools, machines and even habitats, making crewed space exploration and the establishment of off-world settlements throughout the Solar System feasible.

Leaving the System

A lot of critical junctures are on the horizon for our society on Earth, some of which we've already discussed. Perhaps the most fundamental, though, is that after 4 billion years of orbiting our steadfast power source, our Sun, we are on the threshold of mimicking what is happening there. Warmed in the light of our star, we have evolved and explored and innovated, and now we are on the brink of creating our own suns. We can only get them to burn for around a minute, but hopes are high for the public availability of fusion power technology in as little as a few years' time.

Nuclear fusion is a reaction in which two atomic nuclei, typically hydrogen isotopes, combine, in this case forming helium and releasing neutrons, with the difference in mass between the input and output released as energy. Because pressure at the Sun's core is over 100 billion atmospheres, the temperature there – around 15 million degrees Celsius – is in fact less than the temperature required to fuse two hydrogen nuclei here on Earth. Containing this kind of heat has posed a significant challenge. A powerful magnetic field is one way to confine the plasma state of the input fuel, and a tokamak is one of the devices with which to achieve this, typically in the shape of a torus. And now, finally, we are poised on the brink of being able to fuse hydrogen into helium and produce more power than we use to do so. In the few years up to 2024, a series of new temperature and duration records – currently well over 100 million degrees Celsius sustained for around a minute at a time – have been announced by countries including China, Germany, South Korea, the UK and the US. In France, ITER, the collaboration of over thirty countries that will be the world's largest fusion reactor, promises to take our capabilities to higher levels still.

Deuterium and tritium, both heavy isotopes of hydrogen – the most abundant element in the Universe at around three-quarters of all matter – are the fuels currently typically used for nuclear fusion. A few times more efficient even than nuclear fission – with 10 million times more energy available from fusion fuel than from coal, and with fuel readily available, for example from water, and producing just inert helium gas and neutrons as a by-product – fusion promises to revolutionise power generation on Earth. Another kind of fusion fuel is helium-3. Helium-3 is an isotope of helium with two protons and one neutron, and when fused with deuterium may provide more efficient energy production than with tritium, protons being released as a by-product which are more easily contained than neutrons owing to their positive charge. While rare on Earth, helium-3 may be present in much larger amounts on the surface of the Moon; some estimate that 1 million tonnes of accessible helium-3 are available on the lunar surface. Chang'e-1 produced a map of likely helium-3 concentrations in the lunar regolith, and research on both helium-3–deuterium and helium-3–helium-3 reactions, as well as proposals to extract helium-3 on the Moon, are underway.

Mastery of the ways of our Sun which gave birth to us here on Earth may equip us to begin thinking about exploration beyond our home in the Solar System. Fusion propulsion systems could provide efficient and sustained acceleration in space without the need to carry a large fuel supply. One possibility is to fuse helium-3 with deuterium in a reactor, directing the fusion exhaust out of the back of the rocket to provide thrust without the intermediate generation of electricity. Helium-3 is produced by stellar processes, and could be abundant on the surface of a range of celestial bodies beyond our Moon.

We live in a vast Universe containing plenty of the hydrogen, oxygen, carbon, nitrogen and other atoms of which we are made and on which we depend to maintain our state of living. Furthermore, right here in the Asteroid Belt we have conveniently accessible deposits of the less abundant metals and minerals required for the increasing sophistication of our technologies. Extracting resources from asteroids is a more ethical way of mining than disrupting unique ecosystems on Earth, the only place we know of to be teeming with life. Once we have demonstrated asteroid resource extraction and utilisation, and have tested the basic infrastructure required to support human communities on the surface of the relatively harsh environment of our Moon, then it will become possible, driven by the need for new materials, cultural beliefs, wanderlust or otherwise, for us to begin to expand across our Solar System.

With access to different resources in different parts of the System, what about interplanetary trade? A natural currency in an off-world context could be energy, which determines the allocation of resources to all other activities. However, eventually, we may want to move away from centralised notions of value altogether.

Can we imagine a world in which all exchanges of

resources take place for whatever reasons the parties involved feel like? Can we imagine a world of sufficient resources where our objectives as a society advance beyond the material world altogether? A ten-year-old girl listened to a presentation of mine and commented at the end, 'Oh, so living on Mars in lava tubes will be like going back to living in caves, except this time we'll have technologies like 3D printers.' Can we think about going back to the gift economy of our resilient early cave-dwelling human ancestors, but this time with technologically enabled human communities distributed throughout the Solar System and beyond?

Wherever we are in the Universe, even with a far more advanced idea of where we come from and who we are, the one question we will surely still be asking is where we are going. But perhaps the more important question is how: given what we understand about our origins and our nature as humans, what culture will characterise our journey into the new worlds of the future?

8

BACK TO BASICS

The past 4 billion years of life on Earth have culminated in life as we know it today – we stand on the threshold of a new world. We are faced with the greatest challenges and opportunities of our existence: the expansion of society beyond our home planet is within reach, while on Earth the continuous destabilisation of our life-support system has reached a tipping point. We can balance transforming our society on Earth with celebrating our existence in this reality through space exploration; we can live in harmony with each other and the environment, wherever in the Universe we may be. But to do so we will need to go back to basics.

Thinking about living beyond Earth provides us with a powerful thought experiment: if we could establish a society from scratch, what are the fundamental principles on which we would build this new world? We've seen some of these principles alluded to while we imagined a day in the life on Mars; taking an off-world perspective on the current economic system on Earth is a good place to start.

Off-world economics

'Intelligence is based on how efficient a species became at doing the things they need to survive,' said Charles Darwin. The promotion of continuous economic growth on Earth has not always been a good fit with efficiency, nor with intelligence for that matter. Let's examine the notion of economics, its relation to resource management, and how off-world communities may envisage the structure and function of an economic system.

A few years ago I was invited to speak at an event in Tehran. In my time off I visited the Museum of Ancient Iran, along with my compulsory security unit, and was struck by a clay tablet on display. Dated at around 5,000 years old, the tablet was inscribed with Elamite cuneiform text detailing a transaction; including the names of the two parties involved in the exchange, its time and location, as well as the variety and quantity of commodities traded. Each party added their own seal stamp to verify the record. Such records would then be stored, and, fascinatingly, the storeroom secured by a system of pulleys and knotted ropes that prevented undetected access. A system of recording, storing and securing details of transactions is a fundamental aspect of a system of checks and balances of resources: an economic system. While Iran is currently completely disconnected from international payment systems due to US sanctions, it's not without irony that these tablets are intriguing evidence of an early economic system based not on money but on a tamper-resistant record-keeping system.

Dunbar's Number, as defined in the 1990s by anthropologist Robin Dunbar, proposes an upper limit on the number of social relationships a human can manage effectively. Dunbar claimed that we can only keep track of all the relevant social information in a group of about 150 people. In a community where people have a sense of how much each person takes or contributes on average, systems of checks and balances are not necessarily required. Early settlements in what is now Iran were some of the first on Earth to reach population sizes exceeding Dunbar's Number, and were also places where systems of records to keep track of favours and 'I owe yous' emerged thousands of years ago.

As societies grew in size and inter-community trade was established, things like monetary systems and ledgers were introduced to record the exchange of goods and services. Today, with our global communications systems and the amount of time people spend connecting online, our interactive human community has grown to a size of billions. In parallel, our economic system has become increasingly sophisticated over the past few thousand years. Let's have a look at some of the principles.

Earth's current economic system, the first to have global penetration (bar the few remaining uncontacted tribes of people in places like the Amazon or Papua New Guinea), is based on the assumption that individuals will consistently aim to maximise value for themselves.

In this system, value is measured in money. Issues around this fundamental assumption of greed aside for now, do we realise that our money doesn't mean anything? After the First World War, many countries suspended or abandoned the gold standard, which tied currency value to gold, and since 1976, when the US government officially removed all references to gold from their statutes, the value of money has been solely in accordance with a government's declaration of its value. Hence the term 'fiat', which means 'by decree'. The entire monetary system is upheld because we buy into it, literally.

Units of money are basically tokens, representing value with which we can transact to secure basic resources and life-support services among other things. This tokenisation of value, particularly in a system with a fixed number of tokens, enables hoarding, which results in inequality within the system. The lending of tokens as a means of earning more tokens serves to further exacerbate this inequality. Islamic finance has banned such interest charged on loans or deposits as both illegal and unethical. However, it is normal practice across most of the planet. On top of hoarding, we also see the token-rich creating more tokens for themselves at whim. For example, the US central bank, by some counts the most powerful in the world, created approximately 3.3 trillion US dollars in 2020 alone: 20 percent more tokens than were in circulation at the time. Can we understand the fundamentals of what we may require from a system of checks and balances in order to reimagine what an alternative economic system could look like? Data is fundamental to such a system; but are tokens?

An economic system is a means by which societies organise and distribute available physical resources, services and goods across a region – in other words, it's a system of managing resources. An effective resource-management strategy requires knowledge of what resources are around to manage; that is, some kind of quantification of available resources. The current system on Earth does not take the limitations of our finite planet and its biosphere, our life-support system, into account, assuming constant capacity for growth. Unsurprisingly, the system emerged hand in hand with colonisation: the appropriation of a place, its resources and typically also its people by another group with the power and inclination to do so. This assumption of capacity for perpetual growth is problematic for a rapidly growing population with ever-increasing resource requirements dwelling on a single planet; if we all lived like the so-called developed world, we'd need several Earths to support us.

For people living off-world, power and food production and fresh water availability will be monitored with the same fastidiousness that, for example, share price indices are tracked back on Earth. Beyond quantifying availability, a resource-management system could account for transactions in the form of resource transformations, which could be recorded, stored and secured within the system. Is a blockchain-based, secure, connected economic system designed for extreme and resource-constrained environments what we are looking for?

Just months after the financial crash of 2008, which was triggered by increasing deregulation of the industry, the first digital currency to employ cryptography to solve the problem of double-spending without the requirement for a central trusted third-party regulator was proposed. That currency was Bitcoin, now valued at over 1 trillion US dollars. Since then, thousands of other cryptocurrencies have been established. The technology underlying this decentralised capability is a distributed ledger, or blockchain. Transactions are recorded in blocks that are linked and secured by cryptography; these records are verified and stored across a network making the ledger, as well as the rules governing the transactions, resistant to modification.

While adding new information to the Bitcoin network is energy-intensive, high-energy consumption is not intrinsic to blockchain technology in general, and novel consensus mechanisms determining how information is added to the ledger can achieve significant reductions. This combination of capabilities in computing, connectivity and cryptography (the three 'c's) has applications not only in the financial world but also in any transactional environment, including for decentralised data-management systems requiring transparency and immutability. Decentralised systems can withstand multiple failures in the network, and we may therefore consider implementing such systems in areas that we deem critical.

The potential applications of blockchain technology are radical and far-reaching; any system currently under centralised control can in theory be revolutionised by replacing the gatekeepers with decentralised systems of regulation that can be built into this combination of the three 'c's.

One compelling application of blockchain is in securing identity, both online and off. We currently rely on the government of the nation where we are born to provide the systems by which we procure the documentation used to identity ourselves. How to verify, secure and manage identity and personal data online is a major challenge of the current era, not forgetting that as many as a billion people worldwide do not have any formal identification whatsoever. Blockchain can be used to enable individuals to own and control their identities in a decentralised personal data-management system where records are verified and stored across a network, making the ledger resistant to modification. Irrespective of where they happen to have been born on Earth, or on whatever planet or moon they currently reside, with an Internet connection they would be able access the service.

Another application of blockchain is in the management of essential resources. Billions of humans on Earth live without access to adequate shelter and nutrition, clean air and water, and also the reliable power and communications systems that have become important tools to participate in society. Neither governments nor the corporations they increasingly answer to appear to be equipped or inclined to deal with this issue. Inequality and the number of people living in extreme poverty continue to increase, fuelled by the recent pandemic, climate change, conflict and an ever-increasing population of humans confined to a planet no longer able to replenish the natural resources their society consumes. A transparent and tamper-proof system of monitoring global resource availability and consumption, not managed by any particular individual or group of individuals with their own agenda for information-sharing, would be a start towards addressing these issues.

Let's engage in a thought experiment. We're a community of humans in an extreme and resource-constrained environment. This could be the Moon, Mars or even Earth. What resources do we need, and what kind of system do we envisage to manage them? What does off-world economics look like?

Energy is a fundamental resource; to avoid tokenisation, we could make it the currency for the system. Firstly, we'll need to make an inventory of available resources, whether in situ or imported. The value of each item in the inventory is based on critical life-support requirements, coupled with how much energy is needed to produce it. A connected system of sensors monitors all resource levels. An event when a change to any item occurs is logged as a transaction and associated with a cost. Factors influencing cost include the energy required to transform one item to another, as well as the energy required to transform the by-products generated by the transaction back into the resource inventory. Analysing the data contained in such a system of records would enable us to eliminate or minimise by-products that can't be reused – in other words waste, a human invention – while maximising resource-utilisation efficiency with respect to energy consumption.

By recording all resource transformations, or transactions, on a blockchain, we can realise a resource-management system that is decentralised, immutable and transparent: features we would want in an extreme and resource-constrained environment, where organisations of centralised control do not necessarily exist or cannot be relied on, and the survival of the community depends on the secure, transparent and reliable functioning of a life-support system. The data generated, both by the monitoring system and by the life it supports, provides insights for optimisation of strategy during subsequent iterations. And what may these strategies be? In other words, what governance structures may we envisage superimposing on this resource-management system? So as long as continuity of life is at their heart, there are a plethora of options!

While the search for resources drove early human expansion, as our population grew we saw the emergence of doctrine-driven migration. The next wave of migrations may be interplanetary, and seekers of new economic systems may be the participants. But what leaving the planet won't do is allow us to escape from ourselves. Central to the success of any human community are the humans; and, beyond physiological considerations, the psychological and interpersonal factors are among the most complex and critical for success, wherever in the Solar System we may be. Thinking about living beyond Earth gives us the opportunity to reflect on what kind of system we want to build; in other words, if we could establish a society from scratch, what fundamental principles would we want to characterise its culture?

Universal principles

I met cosmonaut Mikhail Kornienko a few months after he returned from a year in space, having spent, along with Scott Kelly, 340 days in Earth orbit in the ISS. I told him about my dream to explore worlds beyond Earth and my plans to move to Mars; I had only recently been selected as one of the Mars100 with the Mars One Project. He laughed, said that I was crazy, but then gave me some interesting advice: 'Go with friends,' he said. While we may one day live distributed through the Solar System or even the Galaxy with a far clearer understanding of where we come from and who we are, one thing already clear is that wherever we are going, we will necessarily be going together.

In many ways, the sophistication of life exceeds our current technological capabilities, and on Mars, or any other off-world environment we plan to inhabit, the collection of species we bring with us will play a critical role in establishing life there. From microbes to humans, living processes will play a central role in the cycling of resources necessary for life to prevail in an environment that would otherwise be deadly for any single species, and certainly any individual organism. But isn't this also the case here? We have seen throughout the history of life on Earth how living organisms themselves create the conditions under which the continuous complexification of life takes place – from the emission of oxygen, which enabled the protective ozone layer and the emergence of multicellular life, to the production of soil, enabling the proliferation of life on land, to the various intricate and symbiotic processes by which all life on this planet thrives together. This interconnectedness is not linear, but rather a vast network of interdependence spreading in all dimensions and in myriad ways through the history of life on our planet. While on Mars the importance of the 'cargo' of life we bring with us may be clearly defined, on Earth we have not collectively acknowledged our reliance on the diversity of life and incorporated it into the way that we live and the infrastructure that we implement to do so. We have plenty of tools at our disposal. It's the way we use our technologies that requires our focus; and specifically our systematic and strategic approach to resource management.

We've been sold the idea that life on Earth is about competition; that evolution is driven by the survival of the fittest. This thinking has played a role in achieving our current level of technological development. However, the question we need to ask ourselves is whether the centralisation of power and resources that has resulted from this competitive ideology is serving our best interests going forward. A far more fundamental natural theme than competition is collaboration: from the cooperative communities that characterised the very first single-celled life forms on Earth to us, with many of the resources we consume produced by the plethora of organisms with which we cohabit the planet. Life beyond Earth is a fragile resource; collaboration and an attitude of appreciation for life and all that is needed to sustain it will likely characterise early off-world resource-management systems. But what about back home?

From archaea to *Homo sapiens*, we all depend on a vast range of fellow occupants of this terrestrial haven for our survival. As humans, more than half of our cells are not human at all but microbes, for example gut bacteria, that we depend on to achieve a range of crucial functions for our survival: without photosynthesis, oxygenation and the emergence of complex life may not have taken place, furthermore photosynthetic biomass provides nutrition for most life on Earth; without fungi, the remains of dead organisms would accumulate and their essential nutrients would not be cycled through food networks, furthermore multicellular life may not have been able to move on to land at all. Collaboration is a fundamental theme in living networks. How can we incorporate community spirit, a common sense of purpose, unity and well-being into a reimagined human culture?

There's a term I came across while doing research into team dynamics in extreme environments: 'salutogenesis', etymologically 'the emergence of well-being'. Sociologist Aaron Antonovsky introduced the concept of salutogenesis while investigating stress management and the maintenance of wellness. Extreme environments do entail stressful conditions, but in spite of, or perhaps because of, the conditions many individuals report a heightened sense of well-being while working in them.

The investigation of salutogenesis in extreme environments will contribute to understanding the protective function of resilience and improved methods of monitoring community health, towards predicting behaviour and developing countermeasures to the effects of tension and stress. However, beyond anticipating things going horribly wrong, more research is required to understand the causes and mechanisms of salutogenesis at both an individual and team level, and furthermore how this heightened sense of well-being may be engendered and maintained in individuals and teams in the extreme environments in which they are living and working. Understanding group-level salutogenesis, the emergence of community spirit in such teams, will have important implications for teams living and working off-world, but importantly also for all human communities in whatever environment they find themselves.

Back to basics

In physics, if we want to understand the behaviour of something, we first 'put it in a box' to understand its characteristics in relative isolation from other things that we are not immediately interested in. What if we removed all pre-existing assumptions about social structures and studied the spontaneous emergence of organisational structures in communities in extreme environments, observed the levels of community spirit that prevail under such conditions, and from multiple data sets drew insights on how we may optimise the way we structure our society, for both on- and off-world applications?

Some of the harsh and remote environments described here are ideal locations for such experiments, where novel integrations of the many existing life-support technologies we have already discussed could lead to the demonstration of resource efficiencies beyond anything we have achieved so far. What about community spirit? We may find that extreme environments often bring out the best in us; inspiring the common vision and unity of purpose that is fundamental to community spirit.

Perhaps. But a more startling realisation is that, as the people of Earth, we are already engaged in such an experiment on a planetary scale. During the past 200,000 years of human existence, our ability to prevail has been tested through dramatic climate-changing events including galactic turbulence, geomagnetic flips, supervolcanic eruptions and asteroid impacts, and more recently through the effect that our ability to shape our environment is having on both our biosphere and our culture. The experiment is not over yet, but one thing is clear: there is no such thing as a single living organism.

South African statesman and philosopher Jan Smuts, who played a key role in the creation of the United Nations, first coined the word 'holism' in his 1926 book *Holism and Evolution*. 'There is something organic and holistic in Nature which shapes her ends and directs her courses,' he said. More generally, holism is the theory that parts of a whole are in intimate interconnection, such that they cannot exist independently of the whole or be understood without reference to the whole, which is thus regarded as greater than the sum of its parts. If we are to advance our understanding of the Cosmos and our place in it, and find a way to transform our society from being defined by the extinction event we are causing on our home planet – the Anthropocene – to one where we flourish in harmony with many extreme and exciting places as we expand beyond Earth, then acknowledging this interconnectedness is a good place to start.

The physicist in me would design an economic system, writing detailed protocols towards a society where people live in harmony with each other and the environment. But in reality, the way in which we create such a society is not from without but from within. Every time we think, speak and act from a place of togetherness, motivated by collaboration and making a contribution to the continuity of all life, we build a new world based on this connection.

Transforming our world

In 2019, I spent a week at (another) silent retreat in Indonesia for my birthday, in between speaking at events in Asia. Before my flight out I went to stay with my friend Julien Mélot, who was designing and building solar-powered boats and lived in an ancient spiritual centre near Denpasar, in between a mosque, a Catholic church and a Balinese temple. Julien told me that sometimes strange things happen there.

That night, I fell into a deep sleep, only to be awakened to teacher Thich Nhat Hanh hovering over me in the lotus position, in glowing orange robes (this was a few years before he passed away in 2022).

'Yes?' he asked, and I faltered; although at the retreat I had been thinking about his teachings after reading a book he wrote, I did not expect a personal visit.

'Um, is my mission to go to Mars?' I asked.

With utter and unparalleled equanimity, the likes of which I have heard about in Buddhist teachings but never experienced to this degree, he replied, utterly free of emotion:

'Mars is there. Mars will be there.' He paused, shimmering above me. He then continued: 'What you seek is within, what is within is what you seek, what you want to know, is what you already know, what you seek is within,' and so on.

He went round in a loop, urging me to rediscover the knowledge I seek within myself. To go out we need to go in. He told me where to start (that's another story for another day, though), before disappearing from my view just as suddenly as he had popped in.

Where do we come from? Who are we? Where are we going? These are some of the most fundamental questions we can ask. Equipped with our curiosity, creativity and ability to shape our world, drawn by inborn wanderlust or otherwise, we have been exploring the unknown since the day we first walked the Earth, out to the edges of the Universe and into the smallest building blocks of reality.

Yet our quest to understand the Universe as something external to ourselves has limited our ability to comprehend it; our methods of probing reality have not acknowledged our unity with it, and we can only see so far under this assumption of separateness. Our deepest scientific enquiries tell us that there is no observer independent of what is observed, no aspect of reality that can be contemplated in isolation from everything else: all energy and matter is connected in this beautiful Cosmos we call home, and nowhere is this more apparent than in the phenomenon of life. The network of all life is intimately connected in time and space on this planet and perhaps beyond. While ancient human culture celebrated this unity, the distinction between ourselves and 'the world' has prevailed through more recent times. Beginning with language, and culminating in our current scientific paradigm and global economic system, we have divided up the world into pieces to try to understand it. This strategy has got us to where we are; look at the fascinating insights into the nature of the Universe, the origins of life, who we are as humans and the exciting journeys that we are now capable of embarking on beyond our home planet as a result.

However, there is conflict between the assumption that we are separate from the world around us and the realisation that we are deeply connected with it: our physics is unable to progress beyond the contradictions between our understanding on the smallest scales and the largest, and remains unable to describe life; our economic system is driving inequality and environmental degradation; while our global society is increasingly synonymous with polarisation and conflict. Our current way of doing things is not sustainable for much longer. We are on the knife edge; what comes next?

Thinking about worlds beyond our home planet has given us a new perspective on who we are, our origins and our future. But while off-world expansion is well within reach, it's not necessary to move to a new planet to build a new world. New worlds begin with us, and the realisation that our bodies are made up of atoms as old as our Universe, molecules older than our Sun, genetics as old as life on Earth, and cells of the community of microbial species helping us to stay alive. A different perspective of the Universe is possible when we realise that we contain within ourselves a record of its entire existence; is this the knowledge that lies within? Whatever planet we are on, the sooner our culture, from our science to our economics, begins to reflect this fundamental interconnectedness, the sooner we can enter a new paradigm centred not just on 'man' but on all life; its environment, its lineage, its well-being, and most importantly its continuity.

Through galactic turbulence, geomagnetic flips, supervolcanic eruptions and asteroid impacts, we have evolved to live with uncertainty. Explorers by nature, we have expanded our society across the surface of our planet. And in the past few decades, we have taken our first steps towards a future among the stars. Some of us are born ready for the journey; we are human and we will keep exploring. Perhaps a more fitting question than where we are heading is how we will go there. Will an us-versus-them mentality continue to drive conflict over resources and a loss of biodiversity which may eventually include us? Or can we find new synergies and symbioses within our human community and, together with the myriad species in our terrestrial network, achieve collaborative goals on this planet and the next, the likes of which we haven't yet imagined?

In the long run, we will have to leave Earth to survive; as we have seen, terrestrial life's time on this planet is 80 percent over. The expansion of life beyond Earth is within reach, but this window of opportunity will not remain open forever. The next few years could see the establishment of the first off-world communities, while at the same time increasingly extreme conditions on Earth are pushing current systems to the edge of their capabilities. This is a moment for reflection on the fundamental principles on which we want to create new worlds. And if we want to build worlds that last, we would do well to learn from the 4-billion-year history of life on Earth: if you want to go far, go together.

EPILOGUE

FOUNDATIONS

A few years ago, I was at an astrobiology conference in Chicago. While being interviewed I introduced myself, saying: 'I'm Adriana Marais and I'm proudly human.' The phrase 'proudly human' came to mind because I truly felt a sense of pride in our humanity, interacting with all the people there. Hundreds of experts gathered, some of whom worked at NASA, others at different space agencies and research facilities all around the world. Some wore beads in their hair, others full suits; but what I realised, standing among them, was that these are the dreamers. These are the kids from the 1960s and 1970s who saw man walking on the Moon and said: 'I want to work in space exploration because anything is possible', united by a wonder for the Universe around us and in particular the life in it.

Meanwhile, the interviewer misunderstood my accent and thought I said, 'I'm partly human', which may be partly true, and certainly an odd remark to have made on record at a gathering of people looking for life beyond Earth! In any case, 'proudly human' had a ring to it; because if not with curiosity, a sense of adventure and foremost a sense of pride in who we are as humans, pride in all who have come before and all who will follow in our footsteps, then with what shall we venture forth?

You may well be wondering if I travelled to so many extreme places just for fun or for some other purpose. Well, both really. I founded Proudly Human, a non-profit volunteer organisation, in 2019, after Mars One declared bankruptcy, with the vision of creating a future we can be proud of, whatever planet we are on. Curiosity-driven space exploration is a celebration of our humanity and the reality in which we find ourselves. Yet the difficulties we face on Earth, including poverty and inequality, mean that not everyone is able to participate. The challenge is to balance ambitious goals that inspire us to dream about, for example, the first human communities on the Moon and Mars with improving conditions for people already living in extreme environments on Earth, and encouraging a rapidly growing young population to get excited about exploring and learning. Proudly Human takes up the challenge!

We have decided that now is the time to launch Proudly Human's Off-World Project. In the midst of the uncertain times we face here on Earth, we aim to demonstrate human resilience, sustainable technology and community spirit in even the most extreme environments through grit, imagination, research and innovation and, most importantly, community spirit at a time when inspiration and unity are needed most.

Proudly Human's Off-World Project is a series of habitation experiments in the most remote and extreme environments on the planet, supported by volunteers, advisors and technology partners. The Project will collect data on groups of experts setting up off-grid infrastructure including shelter, power, water, air, food and communication systems from scratch in the driest deserts, the polar night, as well as under the ocean – where we will aim for a new world record – and live as a research community; to prepare for life beyond Earth, but also to better understand ourselves and our social structures here on Earth. Each experiment will generate exploration-driven innovation and research, and be filmed for people around the world to watch through a documentary series: *Mission Off-World*.

In the meantime, while it's exciting that South Africa will be relaying signals for NASA's return to the Moon, as we did in the Apollo era, it is time for Africa to expand its role in space exploration. What do you get when you combine an economist working to positively impact lives in Africa through space technology with a quantum biologist working on life on Mars? Africa's first Moon mission of course! Economist (and self-described space cadet) Carla Sharpe and I met at an event at the Cape Town Science Centre. Ahead of astronaut Cady Coleman and chief NASA scientist at the time, Ellen Stofan, taking the stage, I was eavesdropping on Carla telling her neighbour that she worked at the Square Kilometre

Array Radio Telescope. The telescope, currently under construction in South Africa (as well as African partner nations and Australia), will be Earth's most advanced radio astronomy instrument, and at the time I was eagerly planning to use the telescope to detect complex prebiotic molecules in space as part of my postdoctoral research. I joined the conversation, and we've been friends and collaborators ever since.

In recent decades, our knowledge of the Universe has been advanced by developments in ground-based as well as space-based observation infrastructure. However, there remains a region hidden from view: Earth's atmosphere, as well as natural and technological radio frequency interference originating from our planet, means that observation of the radio sky at frequencies below 10 megahertz needs to be done from beyond Earth. The Moon is a unique location from which to perform such observations: in addition to the lack of a significant atmosphere, physical shielding on the Moon's farside prevents terrestrial radio interference as well as shielding from the Sun during the lunar night. Low-frequency radio astronomy promises to shed light on the early Universe and phenomena from within the Solar System to the Milky Way Galaxy and beyond not detectable from the surface of the Earth.

Inspired by the challenges of doing low-frequency radio astronomy from Earth and the capabilities of antenna arrays, Carla came up with a design for a simple, remotely implementable radio telescope to do sub-10-megahertz observations from the surface of the Moon. An initiative of the Foundation for Space Development Africa, founded by Carla in 2009 and which I joined as a director in 2017, Africa2Moon is in the running to be Africa's first mission beyond Earth orbit, and the first instrument of its kind to be deployed on the Moon. Achieved through collaboration, Africa2Moon will serve as a continent-wide inspiration and enabler for space-related activities; to educate, encourage and pave the way for African scientists to aspire to achieve world firsts. The Africa2Moon technology demonstrator is currently shortlisted to be launched to the south-pole region of the Moon in 2028; the final payloads are to be announced in April 2025. This could potentially be the world's first lunar radio telescope, to do new science in the sub-10-megahertz frequency range, as well as obtaining critical data for the planning of future (farside) lunar radio astronomy instruments. The full Africa2Moon array is to be deployed and operated on the farside of the Moon, with fifty-four instruments each representing a nation of Africa, the final design to be informed by learnings from the technology demonstrator.

Africa2Moon could be the first African-built technology to perform first-time science on another celestial body. Our simple design aims to demonstrate that with little finance but much tenacity, collaboration and African skill, we can literally reach the Moon and inspire other Africans to reach for the stars. Africa2Moon mission data will be made accessible to researchers across the continent, with the aim of driving increased participation in space exploration and STEM around Africa. And in the meantime, in this exciting era for lunar exploration, we are thrilled that the Foundation for Space Development Africa has recently signed as a partner on China's International Lunar Research Station. When 600 million children in Africa look up at the Moon, we envisage the Foundation contributing to a sense of participation and belonging, in our capabilities as humans, as well as in the Universe, our home.

So in conclusion: watch this space!

INDEX

Index

Index

robotics 18, 122, 204–7
Rockefeller, John 246
rockets 17, 214–19
Rosetta mission 81–2
Ryugu asteroid 82

S
Sabatier process 296, 297
Sagan, Carl 7, 272
salt reactors 256, 290
salutogenesis 294, 326
San people 129–31, 177–8, 179
satellites 215, 218, 224–5, 226–7, 229, 235–7
Saturn 23, 134
  moons 24–5, 46
Saunders, Ben 162–3
scanning tunnelling microscopes 68
Schiaparelli, Giovanni 98
Schrödinger, Erwin 69
Scott, David 189
Scott, Robert Falcon 159–61
sea levels 141, 142, 143, 144
seasons
  Earth 135, 143, 241–2
  Mars 35, 98, 103–4, 275
  Uranus 23
semiconductors 200–1, 248
SETI Institute 109
Shackleton, Ernest 153, 154, 159, 161–2
Shakespeare, William 118
Sharpe, Carla 336–8

shell midden deposits 128–9
Shenzhou spacecraft 219
SHEPHERD 308, 309
Shockley, William 200
silicates 27, 304
silicon 233, 236, 248–9
silicon dioxide 175, 248
Silverstone, Sally 184
Sitchin, Zecharia 21
slavery 146, 147–8
Smuts, Jan 328
Snowball Earth 54–5, 89, 93
solar cells 248–9
solar eclipse 9
solar power 245, 250–1, 290–1
solar sails 14–15, 217–18
Solar System 17–32, 123–5
  formation 43–4
  undulation 136
  see also Inner Solar System; Outer Solar System
solar winds 15, 18, 267
  Mars 36
  Mercury 28
Soyuz 216, 218–19
space debris 227–9, 230, 231, 239
space elevator 215
Space Launch System (SLS) 220
Space Shuttle 216
space stations 225–6, 230
  see also Chinese Space Station; International Space Station (ISS)